Change and Reform in Medicine and Health Education in China

A Teaching Staff's Perspective

Innovation and Change in Education — A Cross-cultural Perspective

Series Editor

Prof. Dr. Xiangyun Du
Aalborg University, Denmark

Nowadays, educational institutions are being challenged when professional competences and expertise become progressively more complex. This is mainly because problems are more technology-bounded, unstable and ill-defined with the involvement of various integrated issues. To solve these problems, it requires interdisciplinary knowledge, collaboration skills, innovative thinking among other competences. In order to facilitate students with the competences expected in professions, educational institutions worldwide are implementing innovations and changes in many aspects.

This book series includes a list of research projects that document innovation and change in education. The topics range from organizational change, curriculum design and innovation, pedagogy development, to the role of teaching staff in the educational change process, and quality issues, among others. A cross-cultural perspective is studied in this book series that includes two layers. First, research contexts in these books include different countries with various educational traditions, systems and societal backgrounds. Second, the impact of professional and institutional cultures such as engineering, medicine and health, and teachers' education are also taken into consideration in these research projects.

For a list of other books in this series, please visit www.riverpublishers.com

Change and Reform in Medicine and Health Education in China

A Teaching Staff's Perspective

Editors

Xiangyun Du

Jiannong Shi

Yuhong Zhao

Baozhi Sun

River Publishers

Aalborg

Published, sold and distributed by:
River Publishers
PO box 1657
Algade 42
9000 Aalborg
Denmark
Tel.: +4536953197

www.riverpublishers.com

ISBN: 978-87-92982-34-6

Contents

1 Change and Reform in Medicine and Health Education — Is There a Need in China? **1**
Xiangyun Du, Jiannong Shi, Yuhong Zhao and Baozhi Sun

1.1 A Global Background for Change and Reform 1
1.2 Change and Reforms in China 4
1.3 Why Investigating the Teaching Staff's Perspective 6
1.4 Defining this Book 7
1.5 Terminology Clarification 8
1.6 Organizations of the Book 9
References 10

2 Educational Change and Reform in Medicine and Health — Globally and in China **13**
Huichun Li, Xiangyun Du and Baozhi Sun

2.1 Change in Medical Education — A Global Context 13
2.1.1 Incentives for Change 14
2.1.2 Changes in Medical Education 16
2.1.3 Reflections and Current Challenges 21
2.2 Change in Medical Education in China 22
2.2.1 Overall Context for Educational Change in China 22
2.2.2 Reforms and Change in Medical Education 24
2.3 Staff Motivation in Educational Change 31
2.4 Summary 34
References 35

3 A Nation-wide Survey on Educational Reform in Medicine and Health Education in China — A Teaching Staff's Perspective **39**

Baozhi Sun, Xiaoju Duan, Jiannong Shi, and Xiangyun Du

3.1 Research Methods 39
3.1.1 Research Background 39
3.1.2 Survey Design 40
3.1.3 Data Generation and Analysis 41
3.1.4 Limitation of the Methods 41
3.2 Respondents' Information 42
3.2.1 Respondents' Backgrounds 42
3.2.2 Gender 42
3.2.3 Age 42
3.2.4 Profession 43
3.2.5 Job Title 44
3.2.6 University Geographical Location 44
3.3 Summary 46

4 Educational Reform — Teaching Staff's Attitudes and Involvement **47**

Xiaoju Duan, Xiangyun Du and Jiannong Shi

4.1 General Opinions on Educational Reform 48
4.2 Which Aspects of the Educational Outcome Should be Further Emphasized 55
4.3 Which Aspects of the Education Should be Improved 60
4.4 Current Inplementation of Curricula Integration 65
4.5 Summary 67

5 Pedagogy Innovation — Teaching Staff's Attitudes and Participation **71**

Xiaoju Duan, Xiangyun Du and Jiannong Shi

5.1 General Opinions on Teaching Methods 72
5.2 Staff's Knowledge About New Teaching Methods 78
5.3 Summary 84

6 Assessment Methods Reform — Teaching Staff's Attitudes and Participation **87**

Xiaoju Duan, Jiannong Shi and Xiangyun Du

6.1 General Opinions on Educational Assessment 88
6.2 Methods to Evaluate Teaching 95
6.3 What Kinds of Evaluation Methods Have Been Used 99
6.4 The Use of Computer-Based Medical Test Database 103
6.5 Universities' Exams and Staff's Familiarity About Them 105
6.6 Summary 113

7 Teaching Staff's Access to Educational Technologies **117**

Xiaoju Duan, Jiannong Shi and Xiangyun Du

7.1 General Opinions on Physical Facilities 118
7.2 Access of Educational Technologies 124
7.3 Expected Educational Technology Training 129
7.4 Access to the Campus Wireless Network 134
7.5 Summary 137

8 Institutional Support for Teaching Staff's Participation in Educational Research and Reform **141**

Xiaoju Duan, Xiangyun Du and Jiannong Shi

8.1 Opinion on the Necessity of a Research Unit 142
8.2 Current Situation of Research Unit and Project Involvement 145
8.3 Research Projects Level 148
8.4 Publication Number 151
8.5 Summary 154

9 Staff's Opinion on General Development of Medicine and Health Education **157**

Xiaoju Duan, Jiannong Shi and Xiangyun Du

9.1 Perception on the General Development of Medicine and Health Education 158
9.2 The Familiarity with New Standards 168
9.3 Summary 172

10 Towards Further Change in Medicine and Health Education in China **175**

Xiangyun Du, Jiannong Shi and Baozhi Sun

10.1 Discussion of Research Results 175

10.1.1 Summary of the Study 175

10.1.2 Teaching Staff's Attitudes Towards Reforms 176

10.1.3 Teaching Staff's Involvement and Participation in Reforms 177

10.1.4 Institutional Supports for Further Reforms 179

10.2 Reflection on Teaching Staff's Motivation and Educational Change 179

10.3 Recommendations for Staff Development 182

10.4 Conclusions, Limitations and Future Perspectives 184

References 186

Appendix **189**

Index **199**

Preface

Worldwide medicine and health education has been an established field of research for over 100 years. Since the mid-19th century, a growing number of educational institutions in medicine and health have established research units focused on conducting research studies in this field. These research units have made a considerable contribution to the improvement and development of educational practice in medicine and health.

In the early 1980s in China, led by the Ministry of Health, China and with the support of the World Bank, a few medical universities established research units focused on medicine and health education. The Research Center for Medical Education at China Medical University (RCME-CMU) was one such research unit. In recent decades, RCME-CMU has conducted a series of research projects at international, national and provincial levels. The topics have included educational reforms and action research on curricula integration, implementation of innovative teaching methods such as Problem-Based Learning, and new assessment methods such as Computer-Based Case Simulation, and Objective Structured Clinical Examination. These projects have contributed significantly to the development of medicine and health educational research in China. Since 2003, these educational reforms and educational research projects have been extended to a wider number of medical universities with the establishment of the North China Center for Medical Education Development (NCCMED).

These projects succeeded in developing and improving educational practices in a large number of medical universities in China. However, the reforms and change processes are experiencing challenges from many fronts such as limited resources, the need to restructure the educational system, and the need to enhance leadership development and staff development. My personal experiences gained through leading and conducting these projects show that one of the greatest challenges is the lack of understanding, awareness, knowledge and skills from the teaching staff. Therefore, in recent years, we have been engaged in conducting projects on staff development for greater pedagogical competence. Currently in China, there is a great need for institutionalized staff development activities and programs, which is one of the most important motivations that has driven us to conduct this research. We expect that the results of this project will make a great contribution to the ongoing establishment of the first institutionalized staff development program in China for medicine and health education.

Finally I would like to express my great gratitude to the China Medical Board (CMB), which gave tremendous support to these projects and the establishment of NCCMED. In particular, I appreciate the support of the president of CMB, Mr. Lincoln Chen, who, in recent decades, has been personally engaged in the development of medicine and health education in China through his passion and dedication. I also hereby express my special gratitude to Dr. Roma Xu at CMB.

I also appreciate the delightful collaboration between Aalborg University and China Medical University in the field of medicine and health education. Our collaboration has helped promote our work on medicine and health educational research in China at a cross-cultural and international level.

Last but not least, I want to thank all the colleagues at RCME-CMU, NCCMED, and Aalborg University who have been engaged in the project. Without their efforts and commitment, this project would not have been carried out and published.

Baozhi Sun
Director of North China Center for Medical Education Development (NCCMED)
December 2012

List of Contributors

Baozhi Sun (孙宝志), MD, Ph.D., Professor, is the former vice President and director of The Research Center of Medical Education, China Medical University, current director of the North China Centre for Medical Education Development, and an honourable doctor at Aalborg University. In the past 20 years, he has led over twenty pioneer projects driving educational reforms, in particular, the change towards PBL in medicine and health education in China. He had also led a few nation-wide projects on accreditation system establishment for quality assurance at Chinese medical universities. He is currently engaged in establishing a first staff/faculty development center for teaching staff in medicine and health education in China. Prof. Sun had more than 100 publications in the field of medicine and health education.

Huichun Li (李会春), Ph.D., is currently a Post-Doc at Aalborg University. He has obtained a bachelor and master degree from School of Education at Beijing Normal University in Beijing, and a Ph.D. degree at Aalborg University, Denmark. His Ph.D. thesis was based on case studies of organizational change towards PBL at a university level in different countries. His research interests include organizational change, staff development, educational innovation and Problem-Based and Project-Based Learning (PBL). He has a series of publications over PBL implementation within various organizational contexts, general education development, and higher education innovation.

Jiannong Shi (施建农), is a professor of psychology working at the Institute of Psychology, Chinese Academy of Sciences and, and also a professor of Graduate University of Chinese Academy of Sciences and a Guest professor of Faculty of Humanity and Faculty of Medicine of Aalborg University,

Denmark. He worked in University of Munich, University of Michigan, Yale University, University of Adelaide, and Regensburg University as a visiting scholar respectively. He focuses his research in the field of giftedness, talent development and creativity theoretically and practically. He and his team conducted a series of cognitive and neuropsychological experiments on cognitive development and brain function of gifted children. Meanwhile, he developed a bio-socio-psycho- model to guide gifted education and creativity cultivation in kindergarten, primary, and secondary schools in China. He authored (or co-authored) 11 books in the field of child development and gifted education, as well as more than 170 journal articles and book chapters since 1990. He is serving as the president of Asia-Pacific Federation on Giftedness of WCGTC and the director of the Research Centre for Supernormal (G/T) Children at the Institute of Psychology, Chinese Academy of Sciences. He is also a committee member of the International Research Association for Talent Development and Excellence and serves as an editor-in-chief of Talent Development and Excellence, the official journal of IRATDE.

Xiangyun Du (杜翔云), Ph.D., is a professor at the Department of Learning and Philosophy and director of the Confucius Institute for Innovation and Learning, Aalborg University. She is also an adjunct professor at Beijing Normal University and at China Medical University. Her main research interests include innovative teaching and learning in education, particularly, Problem-Based and Project-Based Learning methodology in diverse fields such as engineering, medicine and health, and language education, as well as in diverse social, cultural and educational contexts. She has also been engaged in substantial work on pedagogy development in teaching and learning for educational institutions in more than 10 countries. Professor Du has about 100 international publications in relevant research areas including monographs, international journal papers, edited books and book chapters, as well as conference contributions.

Xiaoju Duan (段小菊), Ph.D., is a Post-Doc at the Department of Learning and Philosophy, Aalborg University. She is also an assistant professor at Institute of Psychology, Chinese Academy of Sciences. Her Ph.D. is in educational and developmental psychology. She has published a series of peer-reviewed journal articles in the fields of education, neurosciences, and cognitive development.

Yuhong Zhao (赵玉虹), MD, Ph.D., is dean of School of Continuing Education, vice dean of School of Health Informatics and dean of personnel department, China Medical University. In the past 20 years, she has been actively involved in diverse pioneer projects driving educational reforms in medicine and health education in China, in particular, in the fields of change towards PBL and staff/faculty development. She is currently engaged in establishing a first staff/faculty development center for teaching staff in medicine and health education in China, which is organized by North China Centre for Medical Education Development. Dr. Zhao's publications are focused on continuing education, reforms in health informatics programmes, and overall change in medicine and health education.

1

Change and Reform in Medicine and Health Education — Is There a Need in China?

Xiangyun Du, Jiannong Shi, Yuhong Zhao and Baozhi Sun

1.1 A Global Background for Change and Reform

The medical profession has been under the challenge of transformation. Since the 1980s, different medical associations in the world have called for new medical practitioners of the future (WFME, 1988; WHO, 1991). These calls agree on a list of competencies needed in these new practitioners (Boelen, 1992); they must be able to: (1) assess and improve the quality of care by responding to the patient's total health needs with integrated services, (2) make optimal use of new technologies with ethical considerations, (3) promote a healthy lifestyle by means of effective communication skills, (4) reconcile individual and community health requirements, (5) work efficiently in teams both within the health sector and across the division between the health sector and other related social sectors.

In the 1990s, the Hastings Center in the U.S. launched a project known as 'The goals of medicine: setting new priorities', in which experts from 14 countries were invited to discuss and establish new goals of medicine. A report published in 1996 (Hastings Center Report, 1996; Callahan, 1999) as a result

Change and Reform in Medicine and Health Education in China, 1–12.

of this project proposed four major goals of medicine. These new priorities of medicine include: (1) prevention of disease and injury and the promotion and maintenance of health, (2) relief from pain and suffering caused by maladies, (3) care and cure of those with a malady and the care of those who cannot be cured, (4) avoidance of death and pursuit of a peaceful death. These calls for a new caliber of doctor and new goals for the field of medicine played an influential role in the transformation of Western medicine from disease-centered to patient-centered and the inclusion of health as an essential aspect of medicine (Sun, 2011).

The forces of change in the medical profession are also profoundly affecting the nature of medicine and health education. The history of nation-wide medical education reforms can be traced back to over 100 years ago, when they were impacted by the renowned Flexner report (1910). Worldwide, the current medicine and health education has received increasing amounts of criticism by researchers and professional evaluation organizations due to the incongruence between what is taught at medical school and the actual skills that are needed to provide healthcare service in the profession. Accordingly, an agreement was reached that the approach currently being used to educate doctors must be reformed (The PLoS Medicine Editors, 2005; Aslam, 2006; Cox and Irby, 2006; Cumming and Ross, 2007; Whitcomb, 2007; Lam and Lam, 2009; Cooke *et al.*, 2006; Cooke *et al.*, 2010; Irby *et al.*, 2010; Sun, 2011).

Competency-based education, as a resurgent paradigm in professional education, has been gaining attention in recent years among educators, policy-makers in the healthcare profession, and researchers (Dath and Iobst, 2010). The Competency-Based Medicine Education (CBME) intends to provide an outcome-based approach for the design, implementation, assessment and evaluation of educational programs using an organizing framework of competencies (Frank *et al.*, 2010). It promotes much more of a learner-centered approach, deemphasizes time-based curricular design, and emphasizes the multi-dimensional, dynamic, developmental, and contextual nature of learning. CBME therefore, has significant implications for the planning of medical curricula (ACGME, 2001; Carraccio and Wolfsthal, 2002; Dath and Iobst, 2010; Frank *et al.*, 2010).

A growing number of forces and trends in guiding medicine and health education are focused on competencies as well as educational outcomes.

For example, the World Federation for Medical Education (WFME, 2003) and the Association for Medical Education in Europe (WFME and AMEE, 2005) have established standards for quality improvement at an international level. In particular, in the U.S., the Accreditation Council for Graduate Medical Education (ACGME, 1999) proposed six core competencies for all residents: (1) Patient Care, (2) Medical Knowledge, (3) Practice Based Learning and Improvement, (3) Systems-Based Practice, (4) Professionalism, (5) Interpersonal Skills, and (6) Communication. The core committee of the Institute for International Medical Education (IIME Core Committee, 2002) developed the concept of 'Global Minimum Essential Requirements' (GMER) and defined a set of global minimum learning outcomes that medical school students must demonstrate at graduation. Funded by the European Commission to develop learning outcomes/competencies for degree programs in Europe, the Tuning Project (Medicine) (Cumming and Ross, 2007; 2008) generated learning outcomes for primary medical degree qualifications in Europe. In the U.K., Tomorrow's Doctors (GMC, 2009) sets out the General Medical Council's requirements for the knowledge, skills, attributes, and competencies of undergraduate medical students and for the delivery of teaching, learning, and assessment. These standards provide the framework that U.K. Medical Schools use to design their own detailed curricula and schemes of assessment.

Much of what is found in these new guidelines for education is consistent with the calls for new medical practitioners. Major emphasis is placed on: (1) empathy and respect for patients' well-being, (2) integration and management of illness with health promotion and disease prevention, (3) scientific knowledge with application to medical practice, (4) analytical reasoning, (5) clinical skills, (6) critical thinking, (7) communication skills, (8) collaborative skills to work in multi-professional teams, (9) information management, and (10) professional values and ethics.

The focus on student competencies as outcomes of medical education should have deep implications for curricula contents as well as the educational processes. The aim of medicine and health education, as stated by Sullivan (2005), is to transmit the knowledge, impart the skills, and inculcate the values of the profession in a balanced and integrated way. The traditional and current practice of medical training has been criticized for emphasizing scientific knowledge over biologic understanding, clinical reasoning, practical skills, and the development of professionalism, thereby failing to equip the graduates with the required competencies (The Commonwealth Fund Task Force on Academic Health Centers, 2002; The Blue Ridge Academic Health Group, 2003; Whitcomb, 2007; Cooke *et al.*, 2006; Cooke *et al.*, 2010; Irby *et al.*, 2010). Therefore, reforms are being undertaken in medical schools worldwide, mainly in the aspects of curricula design, pedagogy methods, and assessment methods.

1.2 Change and Reforms in China

Medicine and health education has been of essential importance in China due to the country's large population. This history of formal training of doctors in China can be traced back to 1500 years ago. In the late 19th century, Western medicine and a Western education system were introduced to China, which affected the development of Chinese medical education (Sun, 2011). In the past 60 years, medical education in China has been developing dramatically according to its own needs (Sun, 2011). The major changes are found in the rapid growth of the number of educational institutions and the number of graduates (see Chapter 2). This expansion made a great contribution to the production of medical doctors so as to meet the societal need; however, it also raised issues and problems that limit the quality of education (Ongley, 1989; Deng, 1990; Tu *et al.*, 1994; Gao, *et al.*, 1999; Sun, 2005).

Since the 1980s, reforms were emphasized in medicine and health education in China. These reforms were called for by the Ministry of Education and Ministry of Health. They mainly took place at policy and national system levels, which established guidance in changing the educational practice in all aspects (Sun, 2005; 2011). Since the early 1990s, medical universities in China have been given more autonomy to develop their educational practice according to the local needs, while the government is playing a role of quality assurance (Wang and Lin, 1995).

The current medicine and health education practice in China (see Chapter 2) remains complex and unique (Schwarz *et al.*, 2004; Lam *et al.*, 2006). Facing the contexts of globalization and internationalization as well as the domestic social needs, it is faced by a series of challenges. A recent review work (Sun, 2011) summarized these challenges into the following eight aspects: (1) over-expansion in the students enrollment, (2) limited resources, (3) inadequate number of teaching staff, (4) low quality of clinical education and practice, (5) teacher-centered and lecture-based teaching methods, (6) rigid and inflexible curricula, (7) unsatisfactory graduates who are not able to fulfill the requirements of the profession, (8) poor connection between what has been learned at universities and how it should be practiced at workplace.

Since the 1990s, various innovations have been implemented in the medicine and health education in China, which were greatly influenced by innovation practices in the West, in particular, those in the U.S. (Sun, 2011). In recent years, with inspiration from international standards (Zhou, 2002; Yi, *et al.*, 2004) (as introduced previously in this chapter) and in response to the calls from the newly established national standards (Ministry of Education, 2005), wide ranges of approaches at different levels have been employed to innovate the practice of medicine and health education.

Inspired by Western experiences, a growing number of medical universities have implemented innovations in the aspects of (1) curricula integration (for example, developing clinical skills from the very earliest stages of the study programs), (2) pedagogy innovation for student-centered learning (such as the employment of the Problem-Based Learning method), (3) new assessment methods (such as the Portfolio method, Standard Patients (SP), Computer-Based Case Simulations (CBCS), and Objective Structured Clinical Examination, among others), (4) the employment of educational technologies in the teaching and learning practice, (5) the development of educational research in medicine and health that encourage teaching staff to be engaged in research on students' learning, and evaluation of their own teaching, (6) broadening the contents of medicinal education by including new concepts (such as professional values and ethics, among others).

These innovation practices had a tendency of following international trends, particularly, experiences from the U.S. (Sun, 2011). They led to debates among educational practitioners and health practitioners. Proponents' voices can be heard throughout medicine and health education research conferences, claiming that reforms have helped improve students' skills and motivate teaching staff. At the same time, opponents also publish their doubts concerning whether students' knowledge mastery will be more limited because of the more abridged courses in the integrated curricula (Sun, 2011).

Is there a need to continue reforms in medicine and health education in China? If so, what aspects should be further changed, and in what ways? How should we evaluate the reforms and the application of the innovative practices? These are the questions facing the policy-makers, educators, and practitioners.

1.3 Why Investigating the Teaching Staff's Perspective

Teaching staff has an essential role in the actual conduction of the educational reforms (Cantor *et al.*, 1991; Sheets and Schwenk, 1990; Steinert and Mann, 2006; Steinert *et al.*, 2006). The majority of teaching staff in medicine and health education have been educated as physicians or health professionals instead of educators. They conduct clinical or natural science research and deliver teaching tasks in their professional life at medical universities. In content and time-based systems, their roles (in particular, in the early stages of the study programs) are to transmit knowledge, skills, and attitudes required of a good physician or health professional. Students will develop their clinical skills and better understand professional values during the practice in the later part of education and continual training through residency experiences. The assessment of learning outcomes is in the clinical context using non-standardized and subjective metrics (Carraccio *et al.*, 2002).

In contrast, in the CBME, the role of teaching staff is different. The medical content and core domains are contextual and integrated, and they are taught and assessed based on explicit criteria and standards (Carraccio *et al.*, 2002). The effect of teaching is evaluated by a combination of factors, including management, peers, students, and results of students' performance via standardized assessment. With the CBME being adopted in many countries, the role of teaching at medical universities is becoming more important and at the same time being challenged (Cantor *et al.*, 1991; Mennin and Krackov, 1998; Steinert *et al.*, 2006; Dath and Iobst, 2010).

A series of activities has been initiated in order to engage teaching staff in research and reforms in medicine and health education in China. Supported by the World Bank and Ministry of Health in China, institutionalized medical education research started in the 1980s. A few academic journals were established during the late 1980s and early 1990s. In the past decade, regular annual medical education research conferences were organized by the Association of Medical Education in China. The North China Center for Medical Education Development has also been organizing nationwide annual pedagogy development workshops and seminars for the past three years. However, participants in these conferences and seminars are mainly educational leaders and managers as well as teaching administrators, while actual teaching staff, who are expected to be the major target group of these activities, remain sparse.

Participation in educational reforms can be motivating and, at the same time, challenging to teaching staff due to their dual roles of teaching and research in their own subject fields. In addition, in the Chinese context, the student-teacher ratio remains lower than the average in the Western countries (Sun, 2011). This involves an extra teaching burden falling on the shoulders of the average member of teaching staff. Innovative approaches to teaching and instructing, new methods for assessment, and support of the development of professionalism all require teachers to spend time preparing teaching material, facilitating, and assessing. They also require teaching staff to self-reflect for their own professional development (Cooke *et al.*, 2006).

What are the teaching staff's attitudes towards educational reforms within such a context? What is their involvement and participation in the overall reform practices? How do they perceive the further steps of reform and change in medicine and health education? These questions remain unanswered, and need to be explored.

1.4 Defining this Book

This book is based on a research project on educational reform and change in medicine and health education in China from a teaching staff's perspective. This project is intended to lay a foundation for the establishment of a pedagogy development program (to be established in 2013) in the North China Center for Medical Education Development. The study aims to

- understand teaching staff's attitudes towards the educational reforms that have already been established, as well as their attitudes towards future steps of reforms and change in China,
- gain knowledge about the teaching staff's involvement and participation in the conducted reforms,
- identify needs and contents for planning staff/faculty development activities in China,
- make a contribution to the staff/faculty development practice and research internationally,

- make a contribution to medicine and health education reform and change practice and research internationally.

With these aims, the following research questions are formulated:

- What are teaching staff's attitudes towards educational reforms in medicine and health education in China?
- What are teaching staff's involvement and participation in these reforms?
- What is needed to facilitate teaching staff for further reforms and change in medicine and health education in China?

Supported by the North China Center for Medical Education Development and the China Medical Board, a survey-based investigation was conducted in late 2009 (For the survey, see Appendix 1, and for more information on the survey, see Chapter 3). It addressed 1955 teaching staff members from 22 medical universities in China, with 1820 (93.1%) respondents. This survey was carried out in the form of a questionnaire that consisted of six themes with 58 questions. These themes include the staff's attitudes towards and involvement in the following aspects: (1) general reforms in medicine and health education, such as curricula integration, (2) pedagogy innovation, such as employment of Problem-Based Learning methods, (3) new assessment methods, (4) educational technologies in the teaching practice, (5) institutional support for teaching staff's participation in educational research and reform, (6) further development of medicine and health education. Results of the study are reported and discussed in the remaining chapters of the book.

1.5 Terminology Clarification

Medical and health education: Medical education can be defined as the science and art of preparing future physicians to function properly in society, which should imply responsibility for influencing the circumstances and conditions under which they practice (Boelen, 1992). With a purpose of stressing the importance of health perspective in medical education, the term *medicine and health education* is employed in this book, while *medical education* is the term to be used in citation and references.

Educational institutions for medicine and health: in many countries, the terms such as "medical schools" or "faculty of medicine and/or health" are commonly used to refer to the relevant educational institutions. In China, medical educational institutions were established as separate universities due to the influence of the former Soviet Union's model of education (Schwarz *et al.*, 2004; Sun, 2011). Although a number of medical universities have been merged with other institutions in order to create comprehensive universities since the national level reform in 1999, the majority of medical universities remain detached (Sun, 2011). In the past twenty years, health and technology-related educational programs (such as public health and bio-med engineering, among others) have been gradually established in these universities. In this book, both *medical schools* and *medical universities* are used respectively when referring to international literature and the investigation conducted in China.

Teaching staff: are also known as faculty in the U.S. and staff in U.K. as well as other countries in Europe. *Teaching staff* is the term that is employed in this book, and it is used to refer to teaching administrators, clinical teaching staff, basic studies teaching staff, and public health teaching staff.

1.6 Organizations of the Book

Chapter 2 gives an overview of educational change and reforms internationally and, in particular, in China. This overview includes an introduction to the history of change, drivers for change, and contents and effects of the reform and changes. This is followed by a presentation of related theories concerning educational change, focusing on how change takes place (mainly based on experiences in Western countries) and how such change can be sustained. These theories will be related to the discussion of the research results in later chapters. This chapter aims to establish common ground and set the context for this book.

Chapter 3 presents the methodology of this study — research design as well as data generation and analysis for this nationwide survey on educational reform in medicine and health in China. Also reported in this chapter is the descriptive data about the respondents' information.

Chapters 4–9 report results of this study, structured according to the six themes respectively. Chapter 4 presents findings about teaching staff's attitudes toward and involvement in the educational reforms at the system level, such as curricula integration. Chapter 5 reports results of teaching staff's attitudes toward and participation in pedagogy innovation such as employment of the Problem-Based Learning method. Chapter 6 reports teaching staff's attitudes toward and participation in using new assessment methods. Chapter 7 is about teaching staff's access to educational technologies. Chapter 8 reports institutional support for teaching staff's participation in educational research and reform. Chapter 9 presents results about teaching staff's attitudes toward general development of medicine and health education.

Chapter 10 starts with a summary of the results of the study and discusses these results in relation to relevant theories and contexts introduced in Chapters 2 and 3. This is followed by a reflection upon the overall project as well as its limitations and further perspectives. The chapter ends the book with a conclusion and recommendations for further steps toward reform and change in medicine and health education in China.

References

[1] Accreditation Council for Graduate Medical Education (ACGME). *Outcome project.* Chicago (IL): ACGME. 2001. Available at www.acgme.org/Outcome (Accessed Oct 2012).

[2] Accreditation Council for Graduate Medical Education (ACGME). *ACGME Core Competencies.* 1999. Available at http://www.acgme.org/acgmeweb/ (Accessed Oct 2012).

[3] Aslam, F. Prioritizing Investment in Medical Education. *PLoS Med* 3(3): e159, March, 2006.

[4] Boelen, C., Medical Education Reform: The Need for Global Action. *Acad Med*, Vol. 67, No. 11, November 1992, pp. 745–749.

[5] Callahan, D., *The Goals of Medicine: Setting New Priorities.* Washington, D.C.: Georgetown University Press. 1999.

[6] Carraccio, C. and Wolfsthal, S.D., Englander R, Ferentz K, Martin C. Shifting Paradigms: From Flexner to Competencies. *Acad Med* 77(5), 2002, pp. 361–367.

[7] Cantor, J. C., Cohen, A. B., Barker, D.C., Shuster, A.L., and Reynolds, R.C. Medical Educators' Views on Medical Education Reform. *JAMA.* Vol. 265, No. 8, 1991, pp. 1002–1006.

[8] Cooke M, Irby DM, Sullivan W, Ludmerer K. American medical education 100 years afterthe Flexner Report. *N Engl J Med.* 355, 2006, pp. 1339–1344.

[9] Cooke, M., Irby, D.M., and O'Brien, B.C., *Educating Physician: A Call for Reform of Medical School and Residency.* San Francisco: Jossey-Bass, Carnegie Foundation for the Advancement of Teaching 2010.

[10] Cox, M. and Irby, D.M. A New Series on Medical Education. *N Engl J Med*, 355, 2006, pp. 1375–1376.

[11] Cumming, A. and Ross, M. The Tuning Project for Medicine — learning outcomes for undergraduate medical education in Europe. *Med Teach,* Vol. 29, No. 7, 2007, pp. 636–641.

[12] Cumming, A. and Ross, M. *The Tuning Project (Medicine) — Learning Outcomes/ competences for Undergraduate Medical Education in Europe*. 2008. Edinburgh: The University of Edinburgh. Available at http://www.tuning-medicine.com/pdf/booklet.pdf (Accessed Oct 2012).

[13] Dath, D. and Iobs, W. The importance of faculty development in the transition to competency-based medical education. *Med Teach*, Vol. 32, No. 8, 2010, pp. 683–686.

[14] Deng Y. Development of medical education in China. *Acad Med* 65(8), 1990, pp. 512–514.

[15] Flexner A. *Medical Education in the United States and Canada: A Report to the Carnegie Foundation for the Advancement of Teaching*. Bulletin No. 4. Boston, Mass: Updyke; 1910. Available at http://www.carnegiefoundation.org/sites/default/files/elibrary/Carnegie_Flexner_ Report.pdf (Accessed Oct 2012).

[16] Frank, J. R., Snell, L.S., Cate, O. T., Holmboe, E.S., Carraccio, C., Swing, S.R., Harris, P., Glasgow, N. J., Campbell, C., Dath, D., Harden, R. M., Iobst, W., Long, M. D., Mungroo, R., Richardson, D. L., Sherbino, J., Silver, I., Taber, S., Talbot, M., and Harris, K.A. Competency-based Medical Education: Theory to Practice. *Med Teach*, Vol. 32, No. 8. 2010, pp. 638–645.

[17] Gao T, Shiwaku K, and Fukushima T. Medical education in China for the 21st century. *Med Educ*. 33, 1999, pp. 768–773.

[18] General Medical Council (GMC). *Tomorrow's Doctors — Outcomes and Standards for Undergraduate Medical Education.* 2009. Available at http://www.gmcuk.org/TomorrowsDoctors_2009.pdf_39260971.pdf (Accessed Oct 2012).

[19] Hastings Center Report. *The Goals of Medicine: Setting New Priorities*. Vol. 26, Issue 6, November-December 1996, pp. 1–27.

[20] Institute for International Medical Education (IIME Core Committee). Global minimum essential requirements in medical education. *Med Teach.* 24(2), March, 2002, pp. 130–135.

[21] Irby, D. M., Cooke, M., and O'Brien, B,.C. Calls for Reform of Medical Education by the Carnegie Foundation for the Advancement of Teaching: 1910–2010. *Acad Med,* Vol. 85, No. 2. February, 2010, pp. 220–227.

[22] Lam, TP and Lam, YY. Medical Education Reform: The Asian Experience. *Acad Med.* Vol. 84, No. 9, Sep. 2009, pp. 1313–1317.

[23] Lam,T., Wan, X., and Ip, M.S., Current Perspectives on Medical Education in China. *Med Educ*. 40, 2006, pp. 940–949.

[24] Mennin SP, Krackov SK. Reflections on Relevance, Resistance, and Reform in Medical Education. *Acad Med*, 73(9 Suppl), 1998, pp. 60–64.

[25] Ministry of Education in China (MOE) *The Medical Education Criteria for Undergraduates* 2005 (in Chinese).

[26] Ongley P.A. Asian medical education. *Acad Med* 64 (Suppl. 5), 1989, pp. 22–26.

[27] Schwarz MR., Wojtczak A, and Zhou T. Medical education in China's leading medical schools. *Med Teach.* 26(3), 2004, pp. 215–222.

[28] Sheets KJ. and Schwenk TL. Faculty Development for Family Medicine Educators: An Agenda for Future Activities. *Teach Learn Med* 2, 1990, pp. 141–148.

[29] Steinert Y. and Mann KV. Faculty Development: Principles and Practices. *J Vet Med Educ*, 33(3), 2006, pp. 317–324.

[30] Steinert Y, Mann K, Centeno A, Dolmans D, Spencer J, Gelula M, Prideaux D. A Systematic Review of Faculty Development Initiatives Designed to Improve Teaching Effectiveness in Medical Education: BEME Guide No. 8. *Med Teach* 28(6), 2006, pp. 497–526.

[31] Sullivan W. *Work and Integrity: the Crisis and Promise of Professionalism in America (2nd ed).* San Francisco: Jossey-Bass, 2005.

[32] Sun, B. Z. Trend for Educational Reforms in Medicine and Health. In: *Handbook for Applied Medical Pedagogy*, Sun, Baozhi and Zhao, Yuhong (eds). Beijing: People's Medical Publishing House, 2011, pp. 23–32 (in Chinese).

[33] Sun, B. The New Objectives of Medical Education for the 21st Century and the Reform of Higher Medical Education. *Medical Science and Philosophy*, 26(2), 2005, pp. 24–26 (in Chinese).

[34] The PLoS Medicine Editors, (2005) Improving Health by Investing in Medical Education. *PLoS Med* 2(12): e424, Dec. 2005, p. 1199.

[35] The Commonwealth Fund Task Force on Academic Health Centers. *Training Tomorrow's Doctors: The Medical Education Mission of Academic Health Centers.* New York: The Commonwealth Fund, 2002.

[36] The Blue Ridge Academic Health Group. *Reforming Medical Education: Urgentpriority for Academic Health Center in the New Century.* Atlanta: Robert W. Woodruff Health Sciences Center, 2003.

[37] Tu, M., Xia, X., and Cheng, T. One three-year medical school in China: a reform in Chinese medical education. *Acad Med*, 69(5), 1994, pp. 346–348.

[38] Wang L. and Lin H. The present and the future of higher medical education in China. *Chin Med J*, 108(3), 1995, pp. 163–168.

[39] Whitcomb, M. E. Medical Education Reform: Is It Time for a Modern Flexner Report? *Acad Med,* Vol. 82, No. 1, January, 2007, pp. 1–2.

[40] World Federation for Medical Education and the Association for Medical Education in Europe (WFME and AMEE), in consultation with the Association of Medical Schools in Europe and the World Health Organisation (Europe). *Statement on the Bologna Process and Medical Education.* 2005. Available at http://www.wfme.org (Accessed Oct 2012).

[41] World Federation for Medical Education (WFME). *Basic Medical Education: WFME Global Standards for Quality Improvement.* 2003. Available at http://www.wfme.org (Accessed Oct 2012).

[42] World Federation for Medical Education (WFME). *The Edinburgh Declaration.* World Conference on Medical Education, Edinburgh, Scotland, August 1988. Report, Edinburgh, Scotland: World Federation for Medical Education. 1988.

[43] World Health Organization (WHO). Changing Medical Education: An Agenda for Action. WHO/ED UC. Geneva, Switzerland: WHO, 1991.

[44] Yi, L., Tao, L., and Lu, J. Borrowing medical education standards to promote medical education reform and innovation. *Med Educ* (China) 5, 2004, pp. 3–5 (in Chinese).

[45] Zhou, T. Reflections on applying global medical minimum standards in China. *Med Educ* (China) 4, 2002, pp. 14–15 (in Chinese).

2

Educational Change and Reform in Medicine and Health Education — Globally and in China

Huichun Li, Xiangyun Du and Baozhi Sun

Abstract

Medical education is in a state of perpetual change, and the current trend is directed towards the establishment of a competence-based and student-centered approach in order to meet future societal needs and produce highly competent professionals, both in China and the rest of the world. Specifically, current changes in medical education are centered on several major themes: curriculum integration, pedagogy, use of modern technology, assessment methods, and institutional support. During the change process, the professional bodies, the government, and the educational institutions play different roles in Western and Chinese contexts and therefore formulate different change approaches. Further, both Western and Chinese medical education systems are now encountering various challenges that will be further specified in this chapter.

2.1 Change in Medical Education — A Global Context

Worldwide, medical education is continuously evolving; however, the last century witnessed unprecedented changes in medical education (Enarson and

Change and Reform in Medicine and Health Education in China, 13–38.

Burg, 1992; Papa and Harasym, 1999; Cooke *et al.*, 2006; Lam and Lam, 2009). In the U.S., many nationwide studies and detailed field researches were conducted, which generated a series of influential reports and recommendations producing considerable effect upon the development of medical education in the U.S. and elsewhere. In the early 20th century, the release of the Flexner Report challenged medical education in the U.S. and Canada by criticizing the apprentice system of training in general, in which most medical schools failed to produce competent physicians. It also revealed in a detailed account that quantity had been prioritized over the quality of the education (Flexner, 1910). The report had a remarkable impact in driving an evolutionary change by establishing a single model of medical education that continues to exist, and the recommendations of the report were implemented in accreditation and licensing procedures in North American medical education (Cooke *et al.*, 2006; Irby *et al.*, 2010).

A century later, medical education is facing new challenges in the practice of medicine that the recommendations in the 1910 Flexner Report could not serve to address. Therefore, 100 years later, a follow-up research of the Flexner study was conducted with the same sponsorship to investigate educational innovations at medical schools and residency programs in the U.S. (Irby *et al.*, 2010). Despite different historical contexts, the two investigations addressed similar themes, which were (1) standardization of learning outcomes and individualization of the learning process, (2) integration of formal knowledge and clinical experience, (3) development of habits of inquiry and improvement, and (4) formation of professional identity. They both called for changes and reforms in education to improve the preparation of physicians. Nevertheless, they result in quite different recommendations. Major recommendations from the 2010 Carnegie report can be summarized as: (1) standardizing learning outcomes and individualizing the learning process, (2) promoting multiple forms of integration, (3) incorporating habits of inquiry and improvement, and (4) focusing on the progressive formation of the physicians' professional identity.

2.1.1 Incentives for Change

Many initiatives were conducted by professional groups and boards so as to formulate a guideline for practitioners with regard to medicine and health education. To name a few, these initiatives include: *Tomorrow's Doctors* (General Medical Council, 2009) in the UK, *Learning Outcomes/Competences for Undergraduate Medical Education* (Tuning project) in Europe (Cumming and Ross, 2008), *Global Minimum Essential Requirements in Medical Education* in the US (Institute for International Medical Education), etc. They provided a detailed specification of the key competencies that medical undergraduate students have to develop. The production of these reports was based on intensive collaborative work among the scholars, teachers, students, professionals, and so on, and included large-scale surveys involving different stakeholders in terms of teachers, students, employers, etc. Therefore, these results somehow represented a consensus on the current challenges and the needed competences that could be used to address these challenges. The main aim of these initiatives was to develop a framework to foster doctors and other professionals who would be able to meet the societal needs and address the future challenges in practice. The expected competencies of these initiatives were centered on several common themes in addition to the acquisition of knowledge:

- Clinical skills: to be able to identify medical problems and conditions, perform diagnoses and examination, write prescriptions, manage medical situations and patients, etc.;
- Communication skills: to be able to communicate effectively and appropriately with patients, colleagues, communities, media, etc.;
- Critical thinking and research: to critically formulate hypotheses, collect and evaluate data, etc.;
- Professionalism: professional values, ethical thinking, multicultural awareness.

The specification demonstrates a strong emphasis on the core competencies in terms of performance, communication, and self-learning, contrasting with conventional educational goals, which are primarily concerned with students' ability to memorize and repeat what has been delivered to them. Therefore, the global trend for the change in medical education can be considered as a shift from being knowledge-based to being competency-based. That is, medical education needs to set its focus on students' abilities to take action in terms of dealing with professional situations, communication, and engaging in self-learning activities. This trend also has strong implications for the design, implementation, and evaluation of the medical education and is expected to contribute to the improvement of the quality of medical education. Within the framework of these criteria, medical education should be directed to highlight students' role in learning activities and give them more control over their learning, as well as foster their key competencies.

2.1.2 Changes in Medical Education

In order to facilitate the above-mentioned competencies and better prepare the graduates for their future professions, a growing number of educational institutions worldwide are implementing various reforms and changes. A few essential changes are specified below, which are relevant to the reforms conducted in China and this investigation. These changes include, among others: curricula integration, pedagogy development (innovative teaching methods such as problem-based learning), technology-based assessment methods, and physical facilities (educational technologies).

Curriculum integration

Traditional medical courses are designed on the basis of single disciplines, mainly focusing on the contents and methods from one discipline. The students from this approach are challenged by the fact that they are so specialized that they fail to recognize the connection between different subjects and methods.

Therefore, curriculum integration emerges as a response to the drawbacks of traditional discipline-based curricula and thus aims to facilitate students in drawing connections to the contents and methods from different disciplines in order to acquire an overall, more holistic knowledge. Many different types of curriculum integration models have been developed to date. Speaking broadly, they are referring to either the integration between different basic science courses or between different clinical courses, or even between the two (Qiao, 2011). The Medical School at Harvard University has employed a hybrid integrated curriculum since 1982, which is based on traditional disciplines. It established integration between different basic medical courses, and between basic medical and clinical courses. In contrast, the integrated curriculum initiative at Hawaii began to in 1993, on the basis of the system of the human body. The curriculum integration initiatives bring many benefits for medical education practice. The connection between different subjects is built and consolidated, and therefore students can gain a whole understanding of the practical problem rather than an accumulation of isolated pieces of knowledge. The overlapping content between different subjects is significantly reduced within the integrated curriculum model and thus decreases students' workload. Furthermore, the integrated curriculum is able to facilitate students gaining a deeper understanding of what they are taught by encouraging them to recognize the relationship between different pieces of knowledge and then applying this knowledge to practical situations.

Pedagogy development

Significant changes are also happening in teaching and learning methodology. The emphasis on competency development and the role of the student in the learning process pose huge challenges to traditional didactic approaches in which students simply memorize and reproduce the content conveyed by the teacher. In the last century, many educational innovations have emerged, such as problem-based learning, inquiry-based learning, and action-based learning. Emphasis has been placed on concepts like student-centeredness, practice, reflection, participation, collaborative learning, peer learning, and lifelong learning, among others. Among these methods, the most influential one is the problem-based learning (PBL) approach. Problem-based learning refers to an educational approach where the curriculum content is organized around a real problem situation (Barrows, 1986). Students within the PBL context are first exposed to a medical problem (in contrast to lecture-based learning where students start their learning process by encountering a prescribed body of disciplinary knowledge). Afterward, they work in groups on the problem, investigate the problem situation, identify their learning needs and activities, and eventually solve the medical problem (Boud, 1985). During the learning process, students have the ownership of their learning in terms of identifying their learning objectives, contents, and activities. Generally, it is a student-centered educational approach where students are encouraged to be responsible for their learning, self-direct their learning activities, and develop their competencies. It is expected that, through this method, students will be able to develop their competencies to manage a problem scenario, communicate with other people, and foster both their problem solving skills and lifelong learning attitudes.

The origin of PBL can be traced back to the late 1960s, when the Mac Master University Medical School in Canada introduced a series of simulated patient problems to replace traditional lectures; students worked in small teams on medical problems and the teacher served as a facilitator rather than the knowledge authority directing the students' learning process (Savin-Baden and Major, 2004). The educational strengths of this method have been widely documented and recognized. For example, students of PBL tended to have higher motivation, problem solving skills, group work skills, and self-study skills (Dolmans and Schmidt, 1996; Dochy, Segers, Bossche, and Gijbels, 2000; Bowe, Brian and Cowan, 2004; Strobel and van Barneveld, 2009), when compared with non-PBL counterparts. Currently, PBL has been adopted by a large number of medical schools worldwide, serving as an alternative curricula or replacing the existing curricula entirely, and is reshaping the profile of medical education by moving the focus from teachers' dominant role in learning and disciplinary knowledge to the students' competency development and real-life situations.

Implementing the PBL method calls for improved teaching skills since the role of teaching is different from that in a traditional lecture-based environment. Accordingly, staff/faculty development activities are needed. Maudsley (1999) suggests that staff development should ensure expertise in group processes, raise awareness of the effects of subject knowledge and role modeling, and support tutors who are unfamiliar with the PBL method. More discussion on staff development will be presented later in this chapter.

Assessment methods

Assessment is a powerful driver for the students' learning process. Gibbs (1999) emphasizes that the assessment systems are the most powerful factor influencing students' learning process and that the assessment system can be used strategically in order to enhance students' learning outcomes. The traditional didactic approach is characterized by its strong focus on the acquisition of knowledge content, which does not necessarily enable students to act in practice since there is often a gap between what is known and what can be done. To address students' operational skills, many assessment methods have been developed. OSCE (objective structured clinical examination) has been adopted by an increasing number of medical schools as a way of assessing students' clinical performance. During the OSCE test, students are required to rotate through a series of short stations where they are examined with one or two examiner(s) and either real or simulated patients. Compared with the traditional testing method, OSCE is regarded as highly standardized, and it therefore allows fair comparison between different learners across time and locations. In the last quarter of the 20th century, around 60 percent of American medical schools employed OSCE as a basic method to test students' clinical skills such as communication, clinical examination, medical procedures, exercise prescription, interpretation of results, etc.

Part of the development of OSCE is standardized patient (SP), an individual who is trained to simulate a real patient's symptoms and problems. SP is extensively used in medical and health education to facilitate students to improve their competence in skills such as clinical tasks and communication with a real patient. It establishes an authentic context in which students are allowed to engage in simulated medical activities with a "real" patient. It is convenient since the use of SP does not intrude on real medical work and has fewer ethical problems than working with real patients. Typically, SP is often used as an integral component of OSCE. SP has now been incorporated into the USMLE (United States Medical Licensing Examination), which is mandatory to become a licensed physician in the U.S.

The learning portfolio is an assessment method that has been successfully adopted in PBL context and in medicine and health curricula. It is a rich, flexible document that engages students in continuous, thoughtful analysis of their learning (Zubizarreta, 2004). The portfolio method requires a purposeful collection of student work that exhibits the student's efforts, progress, and achievements in one or more areas of the curriculum (Nikolova and Collis, 1998). This collection should represent the students' best work or best efforts, student-selected samples of work experiences related to outcomes being assessed, and documents demonstrating growth and development towards mastering identified outcomes (Oklahoma, 2001).

In this new era of performance assessment being used to monitor the students' mastery of a core curriculum, portfolios can enhance the assessment process by revealing a range of skills and understandings in each student, support instructional goals, reflect change and growth over a period of time, encourage student, teacher, and parental reflection, and provide for continuity in education from one year to the next (Paulson *et al.*, 1991). Teaching staff can use them for a variety of specific purposes, including (Paulson *et al.*, 1991): (1) Encouraging self-directed learning, (2) Enlarging the view of what is learned, (3) Fostering learning about the learning process itself,

(4) Demonstrating progress toward identified outcomes, (5) Creating an intersection for instruction and assessment, (6) Providing a way for students to value themselves as learner. (7) Offering opportunities for peer-supported growth.

Developed through a process of reflection, evidence, and collaboration, the portfolio may be paper, electronic, or another creative medium (Tynjälä, 1998). At its center, the power of writing and reflection are combined with purposeful, selective collection and assessment of learning endeavors and outcomes in order to improve learning (Dornan, 2002). In practice, learning portfolios have been well-applied into teaching activities in varied learning environments to support reflection, sound assessment, and collaboration. Portfolios have been implemented in a variety of ways, and the application can be at the level of courses, across various disciplines, projects, and programs, etc. It is also a rather contextualized method and the use of it is, to a great extent, dependent on the learning goals, expected learning outcome, motivation level of students, and content of the curriculum, among others. There is an increasing interest in applying it as an assessment method to the PBL environment, due to its strength in promoting reflection and improving process skills in the learning process.

Educational technologies

Educational development is associated with the evolution of modern technologies, especially the rapid development of computer and information science and technology. Traditional educational activities are largely based on the delivery and receiving of textual information. However, for the past three decades, medical education has started integrating an increasing amount of multimedia and interactive elements into educational practice with the assistance of computer technology. The use of educational technology has produced an influential effect on instruction and learning, as well as assessment. Teachers are gradually becoming accustomed to using multimedia slides to deliver the subject content to students. The medical processes and phenomena that reject the demonstration of textual information can be displayed by multimedia sources. Students are using electric portfolios to document their learning process and activities in order to reflect on their learning experiences and prepare the evidence for assessment. *Computer-based Case Simulations* (CCS) has been adopted as a basic method in the USMLE in the U.S. since 1999 and it formulates an authentic environment in which students are exposed to a complex and interactive simulated medical situation. In general, the new technological improvements in education benefit medical education from various points of view; they can deliver medical information and processes which cannot be displayed through textual information, they can create a highly authentic environment in which students can engage in simulated medical situations, and they can be adopted in assessment so as to test students' competency development.

2.1.3 Reflections and Current Challenges

The change of medical education so far is primarily intended to address two major concerns. First, medical education is expected to be able to meet the increasingly changing social needs, as noted in the set of criteria in *Learning Outcomes/Competences for Undergraduate Medical Education* in Europe, which proposed a set of requirements that future clinicians and doctors have to satisfy.

Further, in order to foster the future generation of doctors, the educational system requires a transformation with regards to its objectives, instructional approaches, assessment methods, and the like. This transformation should occur in such a way that it should emphasize the role of learner, skill development, core competencies, interaction with others, and the ability to make ethical reflections.

It is worth noting that although there are abundant recommendations for the improvement of medical education, they are not necessarily put into practice (Enarson and Burg, 1992). The reasons may be due to the nature of the research; the nationwide research seems to successfully put forward general suggestions which may lead to national educational change, but they do not pay sufficient attention to the local context of a specific organization, and therefore, the recommendations seldom result in an organizational change.

In contrast, detailed field work is able to generate information in a particular context and could therefore offer deeper insights into how to introduce educational innovation within a specific context. Therefore, in order to facilitate educational innovation at a medical school, more field work concerning the local context is needed.

Further, Enarson and Burg (1992) link the failure to induce large educational reform to the organizational characteristics of modern universities in which education becomes increasingly inferior to the needs of research and clinical missions of the universities. Since the policy regarding the staff's income and promotion is largely dependent on the staff's performance on researching and clinical work, the staff lacks the incentive to participate in the reform of medical education.

Moreover, the traditional didactic approach and culture still dominate many medical schools in which the main aim of education is to memorize the knowledge content delivered by the teacher. Teachers play a dominant role in the teaching and learning process. They design the learning objectives, select and arrange learning material, and determine the assessment method. The assessment method is also concerned with students' quality of knowledge retention. The favor of the traditional approach makes it quite difficult to initiate educational innovation or sustain the results of the educational change.

2.2 Change in Medical Education in China

2.2.1 Overall Context for Educational Change in China

Education is in a state of constant change. Traditionally, Chinese education is aimed to cultivate one's personal character at an individual level, and to foster social servants or managers at the societal level. For a considerable period of time after the foundation of People's Republic of China in 1949, the education policy aimed to foster the students' political loyalty to the country. All educational institutions, from primary school to higher education institutions (see Appendix 2), are completely financed by the national and the provincial governments. The educational objectives, the curriculum design, and the knowledge content of education are largely guided by the government. The educational approaches are largely dependent upon the teachers' dominance of the class and the learning process. Since the 1970s, education in China has undergone drastic changes regarding educational objectives, administrative mechanism, financing, curricula, instructional approaches, and so on.

There have been several themes regarding educational change in the past three decades in China. The first theme was to shift education from test-directed to quality-oriented. Traditional educational approaches are always termed as test-oriented education since they are primarily concerned with preparing students for the examinations. Within this context, both teachers and students are concentrating on knowledge retention and students' academic performance in the test. However, these approaches have difficulties in equipping students with skills.

It was for this reason that, in the 1990s, the government launched national initiatives to shift the educational focus from preparing students for tests to fostering students' skill development. These initiatives are termed as quality education, which aim to enhance students' skills and competencies, rather than merely facilitating their ability to memorize the knowledge content.

Another drawback of traditional higher education in China is that it is far too specialized, implying that students are trained to concentrate too much on their own disciplines. Therefore, students' horizons are too narrow, and they know very little about the world outside their own discipline. Further, students are able to see neither the connection between different subjects, nor the connection between theories and practice.

The first concern led to the national initiative of promoting humanity quality education or general education, which is intended to broaden students' knowledge base and facilitate their understanding' of other fields' theories and methodologies. For example, students of natural sciences are required to enroll in some Humanities courses, and in turn, students in Humanities and Social Sciences are expected to study something from the field of natural sciences. To address the second concern, some disciplines are encouraged to transform their single discipline-based courses to "integrated courses" so as to change students' learning content from fragmented disciplinary knowledge to the whole. For example, in the medical domain, many medical education institutions have integrated their traditional single-discipline based subjects to "integrated courses."

The educational approaches are also in the process of change. The pedagogical changes allow teachers to employ various types of methods to motivate students and enhance their teaching effectiveness. These pedagogical innovations include action-based learning, research-based learning, and inquiry-based learning, among others. Students are encouraged to take more responsibilities for the learning process.

Another important dimension is the change of the management system. Traditionally, all educational institutions are funded and owned by the government. Since the 1980s, educational institutions have been allowed more freedom to manage their own affairs, although they are still largely influenced by the governmental agenda. Further, private investment is encouraged in education, which leads to tuition charges, establishment of private educational institutions, and so on. These changes are deeply shaping the educational profiles in China in terms of educational objectives, curriculum, instructional approaches, school-society relationship, teacher-student relationship, etc.

2.2.2 Reforms and Change in Medical Education

An overview of educational change to medical education in China

Chinese medical education has a history of over 1500 years; however, the current medical educational system is largely influenced by the Western medical educational system. Since the foundation of People's Republic of China in 1949, China has established a multiple-level medical educational system, covering secondary medical education, higher medical education, and graduate and continuing medical education. The current system for training medical students is quite complex, with a duration of 3 years for secondary medical school and 5–10 years for tertiary medical education (Regarding medical higher education in China, see Appendix 3). In 2008, China had 184 medical universities, with 1,650,000 on-campus students and 400,000 graduates. Since the 1970s, medical education has seen several major reforms regarding structure, knowledge content, learning objectives, and processes, as well as assessment methods.

One important change is the establishment of the national accreditation criteria system for medical undergraduate education. The motivation behind the launching of this national initiative was to bring Chinese medical education system in line with international standards. In 2001, the World Federation for Medical Education released the Global Standards for Basic Medical Education, which exerted a large influence on Chinese medical education. Later, the Ministry of Education organized several meetings to discuss these standards and assigned the Research Task force of Quality Assurance for Chinese Medical Education to localize these global initiatives and develop new standards to guide the future direction of Chinese medical education. In part, national initiatives were aimed at establishing a nationwide accreditation criteria system to guarantee the quality of medical education in China. In 2005, the Ministry of Education released the Medical Education Criteria for Undergraduates, which roughly specified three major learning objectives for medical undergraduates in terms of moral and career quality, knowledge coverage, and skill development. Compared with traditional medical education, which is primarily concerned with medical knowledge retention, this report proposes social, ethical, and action dimensions that learners need to address. For example, students should have an awareness of medical ethics in practice and respect patients' confidentiality, respect the patients' personal religion and understand their personal and cultural background, be able to communicate effectively with patients and their relatives, and be able to educate patients and others regarding a healthy lifestyle and disease prevention. Since these accreditation standards partially address social needs, they also, to some extent, direct Chinese medical education to become more adaptable to social needs as an increasing number of medical schools join this accreditation system. The development of medical education in China is largely prompted by this national initiative.

The second major change is that new knowledge and new social concerns are being integrated into medical education. In the last century, there was a conceptual change to the dominant "Diagnose-Treat" model in medical education. The conception of this model is primarily concerned with identifying the biological causes of disease and then taking measures to treat it. However, with the development of medical practice, it became gradually recognized that not all diseases can be identified with a cause.

Furthermore, the "Diagnose-Treat" model tends to reduce patients to merely objective carriers of disease to be manipulated and thus ignore the personal, social, and cultural contexts of patients. As time goes on, the importance of the patients as human beings has been more widely recognized, and this fact mandates that medical practice address the relationship between doctors and patients. This new knowledge and these new social concerns have gradually become an integral part of medical education at higher education institutions, leading to an increasing number of medical schools starting to set up new courses on healthcare, the doctor-patient relationship, medical ethics, etc.

The third major change is the reform of course integration. Since the inception in 1949, Chinese higher education followed approaches from the Soviet Union, which was highly specialized. Although it was beneficial in its time in that it produced urgently needed professionals for society, the approach also faced huge challenges when it was found that the students were becoming too focused on their own disciplines and were failing to recognize the relationship between different disciplines. They also failed to see the relationship between their subjects and the society at large. In other words, the highly specialized educational model hampers students from formulating a broader perspective.

Furthermore, from the perspective of problem solving, since problems in modern society are becoming increasing complex and cannot be readily solved by an approach based on one single discipline, it has become necessary to direct students to formulate an interdisciplinary perspective. At the national level, such incentives led to initiatives of merging traditional disciplines, and consequently resulted in the reduction of disciplinary numbers from 57 in 1987 to 20 in 1998 in medical education domain in China. At the organizational level, there was a general trend towards merging disciplines, which was firstly confirmed and advocated in the Medical Education Criteria for Undergraduates in 2005 by stating "the medical school should take an active role in integrating courses and the knowledge content in a horizontal or vertical manner." Under this encouragement, a large number of medical schools joined this initiative and took measures to integrate their courses and knowledge content. These integrations included mergers between basic courses, between clinical courses, and between basic and clinical courses. This initiative eventually led to a significant reduction of the number of medical courses. For example, from 1998 to 2001, the number of courses in China Medical University was reduced from 17 to 9.

Another major reform focused on the teaching and learning process in medical education. Traditionally, lecture-based learning is the predominantly used method in Chinese medical educational practice. Teachers make presentations to students and deliver the prescribed body of knowledge content to them. The learning process is mostly controlled and dominated by teachers. The educational objectives are primarily concerned with the acquisition of knowledge. The assessment is also aimed at testing the quality of students' knowledge retention. However, when the educational objectives are shifted toward promoting students' competence in taking action, communication, and dealing with the doctor-patient relationship, the teaching and learning process should also be transformed to allow students to take a more active role in the learning process.

The Medical Education Criteria for Undergraduates encouraged all medical education institutions to carry out educational reforms characterized by student centeredness and self-learning (Ministry of Education, 2005). Regarding the reform of the teaching and learning process, many medical schools are engaged in a process of transforming their lecture-based learning approach to the (Problem Based Learning) approach, or at least integrating some PBL elements into their educational practice. Many medical education institutions, such as Wuhan University (Wang, Tai and Huang, 2008), Hong Kong University, China Medical University, etc., are currently innovating their educational approaches so as to enhance their teaching and learning quality and cultivate high quality medical professionals. With this approach, the learning process begins by presenting students with a series of medical symptoms rather than simply imparting them disciplinary knowledge. Following this, an analysis of these symptoms will generate a number of problems that students need to address. Afterwards, students work on these problems in groups. By investigating medical problems in context, students are expected to acquire the intended body of knowledge. In contrast to lecture-based learning, the PBL learning process is directed by the students in that they control their own learning process. Teachers within this context are mostly serving as facilitators to support students rather than instruct them. In addition, since students are working on the real medical problems in their groups, not theoretical knowledge, they are better able to see the connection between the theory and the practical life. It is also worth noting that there is no single PBL definition in practice, PBL curriculum model could be quite diverse (for example, see Appendix 4).

Associated with the reform of the curriculum content and the teaching and learning method is the change of assessment method within medical education. Traditionally, the assessment method is primarily concerned with students' acquisition of knowledge content from the lecture. However, as learning objectives began to integrate students' skill development (in terms of problem solving skills, communication skills, and group skills), it was felt that their social, cultural, and ethical awareness growth were also an integral part of their learning outcome. Therefore, the assessment methods should be altered to reflect the change of the educational objectives. When compared with the change of focus in the teaching and learning methodology, the change of assessment method is relatively low-scale.

What happens now is that formative assessment has gained wide recognition. Traditionally, the assessment was largely focusing on students' performance in a final examination. Currently, however, at many universities, the assessment begins to concern students' daily performance such as attendance, participation in group work, or classroom discussion. An assessment method focusing on the learning process rather than the learning product is widely believed to be more beneficial for learning. It also points out a new way of seeing assessment, changing it from "assessment of learning" to "assessment for learning." Individual innovations to assessment are also reported. For example, the Medical School at Shan Tou University introduced OSCE (Objective Structured Clinical Examination) in 2004 to test students' clinical competence through the use of a simulated environment and standardized patients, which proved to be quite promising (Zheng and Yang, 2010). However, the systematic change of the assessment is still challenging, since assessment systems at a higher education institution is entangled with other elements which are quite difficult to transform, such as governmental and organizational policy, and students' motivation to seek for academic awards and job opportunities.

Incentives for medical educational change in China

The changes occurring in medical education are firstly due to the incentive to meet the needs of society. In the medical practice domain, it is worth noting that several major changes have taken place. For example, medical science has concerned more with health care, disease prevention, and maintenance rather than merely offering remedies for the disease. Furthermore, medical practice involves dealing with patients, so patients' social, cultural and ethical concerns should be taken into consideration in medical practice. In this regard, new competencies are demanded, such as communication skills, ethical, and social awareness, problem solving skills, etc. In China, although medical schools traditionally do not have an intimate connection with the labor market, they should also take the social requirements of high graduate competence into account when designing their learning objectives and content.

Another major incentive to make educational change in medical education is based upon the motivation to promote educational quality. In China, medical education has been suffering from a great many problems, such as poor skill development, a collective narrow-minded mentality, lack of social, cultural, and ethical awareness, low learning motivation, general inertia in terms of taking action for change, etc., Therefore, the educational system has to make innovations regarding to its curriculum, learning content, teaching and learning method, assessment, and so on. One of the significant changes is concerned with the teaching and learning method. As we noted earlier, the PBL approach has been employed as an alternative educational approach by an increasing number of medical schools.

The role of government in medical educational change

In China, most of the higher education institutions are funded and administrated by the government, and therefore, are subject to governmental policy and requirement. Within this context, governmental initiatives play a vital role in triggering the changes in the educational field. It can be noted from the above description that the establishment of the accreditation system, course integration, and reform of teaching and learning elements are primarily initiated by governmental policy. Under the guidance and the encouragement of the government, medical schools begin to make changes so as to meet the governmental requirements. Indeed, these reforms gradually change the theories and practices in the medical field, and significantly shape the profile of Chinese medical education, though traditional lecture-based learning still has huge influence. As medical schools gain increasing levels of freedom from the government and are allowed making their own innovations and explorations, they will play a more active role in leading educational changes.

Current situation — challenges and barriers

Increase in student numbers. Since the late 1990s, Chinese higher education in general has witnessed a drastic increase in student enrollment number due to the national endeavor to make higher education more accessible for the public. Within this context, medical education in China has also seen a significant increase in the number of students. In 1998, the medical higher education institutions had an enrollment of around 82,500 students. However, the enrollment number of students in medical higher education institutions had increased to 454,400 by 2007. The graduates from medical school equally increased from 148,000 students in 2003, to 1,262,200 in 2008. The abrupt increase in student numbers in medical education posed significant challenges to both educational theorists and practitioners. For example, alongside the expansion of the higher education scale, medical education institutions also recruit an increasing number of staff members, who are, to some extent, lacking the experience in performing teaching activities and reflecting on their educational behaviors.

Resource limitation. Two aspects highlight the resource limitation. Firstly, due to the fact that China is a large country and that it has developed in a somewhat imbalanced way, the economic gap between different regions, organizations, communities is considerably large. Both economic and human resources are relatively centered on the Eastern, urban area rather than the Western, rural one. Therefore, the medical schools in rural Western areas of China are more likely to have fewer resources than those in Eastern, urban areas, and their development is largely restrained by resource limitation. Further, the massification of higher education also poses significant challenges to the resources of medical schools in China. Many medical schools are low in staff, library books, electronic resources, classrooms, and dormitories, considering the large number of students enrolled. More importantly, the resource limitation also restricts student access to modern technologies such as multimedia and SPs in many areas.

Traditional education. Current educational practice in the medical domain is heavily dependent on traditional educational approaches. The flaws of traditional educational approaches are disclosed. From the perspective of curriculum design, traditional curricula are centered on isolated subjects, which focus on the delivery of knowledge content rather than the fostering of skills. Besides, there is also a lack of integration between different subjects. Students may perform in a satisfactory manner when memorizing the given body of knowledge content, but are still unable to see the connection between the theory and the practice. Students are weak in terms of skill development, lacking in problem solving skills, communication skills, teamwork skills, and so on. Students are not especially motivated and engaged in learning since the educational approaches do not interest them. Furthermore, since educational practitioners still highlight the value of knowledge memorization to education, the assessment is unlikely to change, and continues to emphasise knowledge retention rather than competency development. Since the assessment method may define which methods students are likely to employ in their learning process, an assessment method focusing on knowledge retention may direct students to rote learning and further undermine the effort of making pedagogical change.

Traditional philosophy. Apart from the influence of the traditional education paradigm, one challenge that might hinder the establishment of a student directed learning paradigm is the traditional Confucius-based mindset, which places the teacher as a superior over the student in both a cognitive and ethical regard. The cognitive superior refers to the fact that teachers are, by nature, more knowledgeable than students, and therefore, they should guide students in the whole learning process so as to guarantee the correctness of the students' learning outcome. The ethical superior is more radical in Chinese culture and it comprises one of the basic unwritten rules guiding the relationship between teachers and students. With this ethical concept as a foundation, students are expected to conform to their teachers and follow their instructions. Within this tradition, students are more likely to respect their teachers rather than challenge them in a direct manner. This ethical tradition is quite inconsistent with a student-directed education paradigm, where an equal and communicative relationship between student and teacher is required. The historic tradition of the didactic approach, alongside the aforementioned Chinese ethics, considerably hampers the formulation of a student-centered approach.

2.3 Staff Motivation in Educational Change

Ways of implementing educational changes vary. Two commonly seen strategies to initiate a change are top-down and bottom-up approaches (Kolmos and De Graaff, 2007). Top-down approaches refer to the change mainly initiated from the top management level, and they work effectively in dealing with the emergent needs. However, top-down changes are likely to meet difficulties since they sometimes lack the support from the bottom level. Bottom-up changes are heavily dependent on the bottom levels' enthusiasm and participation. Although they gain the support from the bottom level, they cannot last long without the support from the top level. Thus, in an ideal situation, educational changes should enjoy both managerial support and the recognition of the need for change from the bottom level.

Fullan (2001) specifies the possible factors that could influence the implementation of educational change. Among all the factors, a sense of urgency should be established prior to the change (Cohen, 2005). The organizational members are not likely to engage in the change process if they are not aware of the urgency of the current situation and the need for making improvement; therefore, it is necessary to help them to recognize the organization's needs and drawbacks and the gap between the organization's current situation and its intended potentials, so as to help them to see the necessity of change.

The motivation of the organizations' members is crucial for change, since it is the source of momentum for the change. As noted by Kotter (1995), Moesby (2004), and Cohen (2005), the maintenance of the momentum in the organization is always the key factor for a successful organizational change. Any change which lacks the motivation of the organizational members is unable to sustain itself in the long run. Since members' motivation is such a valuable key factor, various methods have been developed to manage the change and motivate the organizational members to participate in the relevant activities.

In general, there are three types of approaches stimulating the organizations' members: rational, power-coercive, and the normative-re-educative approach (Chin and Benne, 1985).

The rational approach is to inform the members of the benefits of the change. As long as the members recognize the personal gain, such as economic reward, they are likely to be more eager to join the change initiative.

The power-coercive approach is more concerned with executing the managerial intention, regardless of the willingness and the perception of the organizations' members. In general, this change strategy is mainly initiated from the top and could be quite effective in dealing with emergent problems. However, without momentum from the bottom, the change is unsustainable.

Therefore, the third approach, the "normative-re-educative" approach is essential since it is the approach most likely to lead to a change in the members' mindset and willingness to engage with the change program. This approach can take various forms, such as workshops, seminars, conferences, etc., to impart the necessary knowledge and skills to the organizations' members so that they become cognitively and, hopefully, emotionally motivated to participate in the change.

Moreover, a normative approach, which is primarily concerned with aligning organizational members' conceptions of change with managerial intention, is far from being sufficient. Within this regard, a hermeneutic approach is necessary so as to get a genuine understanding of how organizational members interpret change in reality (Henriksen *et al.*, 2004). The hermeneutic approach is not to design a technique to direct staff motivation and behavior; rather, it is more concerned with disclosing how organizational members understand and interpret the organizational phenomenon and process. This approach is quite essential for educational change since it can provide a deep insight into the worldviews of the organization's members regarding change. In order to implement a successful change, a hermeneutic approach should be adopted alongside the use of a normative approach.

In relation to educational change in medicine and health education, staff/faculty development plays an important role in the change process in pedagogy, didactics, and management (Dolmans, Wolfhagen and Snellen-Balendong, 1994). The value of staff development programs with respect to educational innovation has been widely recognized (Dath and Iobst, 2010), since they can serve to promote the motivation of the staff in the educational change process. The incentive of the staff is a major factor for gaining momentum for the change. Technologically, the aim of the staff development program is to equip staff members with relevant knowledge and skills regarding the intended educational change. As for the development of competency-based medical education, staff members could be informed of the key competencies that students have to develop and the teaching techniques in different domains of medical education, as well as new strategies to design authentic and regular assessment. Additionally, it is expected that through an educative process in terms of seminars, workshops conferences, etc., conceptual change could also happen to staff members so as to enhance their recognition of the importance of change and eventually formulate a consolidated base for the change. Therefore, it is quite necessary to discern a way to support teaching staff in gaining the knowledge and skills and enable them to get through the personal development process for the educational transformation (Maudsley, 1999; Savin-Baden, 2003).

In Europe, staff development programs received adequate attention during the 1990s. A number of staff development centers have been established in some universities. Training activities for improving pedagogical skills are offered to assistant professors at many universities in Northern Europe. Participation in pedagogical training is required for newly employed teaching staff due to a formal requirement of individual pedagogical qualifications. This established a platform for educating staff, especially new and young staff members, with regard to new teaching and learning methods. In general, documentation of the effectiveness of staff development remains sparse. One study has, however, shown that the long-term impact of staff development activities depends on contextual aspects, including the extent to which the staff development is integrated into the work setting of the participants (Stes *et al.*, 2007). In the U.S., staff development also continues to be a task full of challenges (Brent and Felder, 2003). It has been a strenuous task to invite engineering professors to participate in training activities to improve teaching skills, partly due to their mistrust of the effectiveness of the training and partly due to lack of time and motivation (Brent and Felder, 2001). A systematic review of the effectiveness of faculty/staff development initiatives from 1980 to 2002 demonstrates significant positive changes in staff members' attitudes, knowledge, and skills, as well as teaching behaviors (Steinert *et al.*, 2006). Experiences in staff/faculty development so far also show that there is a great need for an institutionalized support system to organize sustainable activities in order to help establish educational innovations and reforms in the long term.

2.4 Summary

Over the past few centuries, medical education has seen constant changes, both in China and globally. These changes are characterized in two striking ways. Firstly, medical education is becoming increasingly intertwined with societal needs. Medical education is not simply concerned with the traditional cognitive development of students, but rather considers the social needs an integral part of educational objectives. Medical education is no longer aimed at merely transmitting knowledge and skills to students; rather, it requires students to be able to handle the social relationships with patients, their colleagues, etc. Furthermore, medical education itself has become more concerned with the improvement of educational quality. The endeavor to consolidate a research basis for medical education, to establish a more student-centered teaching and learning approach, to encourage group work, etc., is aimed at establishing a higher quality medical education system.

It is worth pointing out that medical professional bodies are essential in assessing the educational quality, setting up educational objectives, and proposing future directions for the development of medical education. The release of the *Flexner Report*, *Tomorrow's Doctors*, was a milestone and had a significant influence on the development of medical education.

One major difference that distinguishes medical education in China from many Western countries is that national initiatives play a vital role for the change movements of medical education. During the past several decades, most of the major educational changes were initiated and directed from governmental level, such as the movements from test-oriented education to quality education, massification of higher education, humanity quality education, etc. These movements are initiated and approved by the national government, and later expanded to higher education institutions. Medical education is important to China's large population. The undergraduate medical education system is being streamlined and national standards are being established. Innovations in medical education have recently been encouraged and supported, including the adoption of PBL. In recent years, there were increasing organizational initiatives at the university level, such as the PBL implementation at Wuhan University and China Medical University. These university initiatives may imply that future educational change will involve many more change agents than simply the government's initiatives.

Regarding challenges for medical educational change, although medical education institutions globally encounter the challenges brought by the traditional educational thoughts, which largely influence teachers and students' value and action, some issues are quite specific to Chinese medical education. The traditional didactic approach still dominates the current educational practice and largely hinders the formulation of a student-centered and competency-based educational approach. The didactic tradition, to some extent, obstructs staff members from seeing the current societal needs and realizing the urgency for making educational change. It may also be responsible for the reluctance to make changes to the assessment method, which is still largely based on testing students' knowledge acquisition. The traditional Confucius-style mindset, which guides how to deal with the teacher-student relationship, may undermine the effort to build up a student-centered teaching and learning approach. The expansion of higher education, coupled with the imbalanced development in different areas, intensifies the lack of resources for many medical schools. The access to modern educational technologies and the staff members, infrastructure, and facilities are largely confined by the limited resources, which further leads to difficulties in making educational improvements in terms of new pedagogical approaches and assessment methods.

References

[1] Barrows, H.S., "A Taxonomy of Problem-based Learning Methods," *Medical Education*, Vol. 20, No. 6, 1986, pp. 481–486.

[2] Bloom, S., "Structure and Ideology in Medical Education: An Analysis of Resistance to Change,"*Journal of Health and Social Behavior*, Vol. 29, December 1988, pp. 294–306.

[3] Boud, D., *Problem-based Learning in Education for the Professions*. Sydney: Higher Education Research and Development Society of Australasia, 1985.

[4] Bowe, B. and Cowan, J., A Comparative Evaluation of Problem Based Learning in Physics: A Lecture Based Course and A Problem Based Course. In Savin-Baden, Maggi (Eds.). *Challenging Research in Problem Based Learning*. Berkshire: McGraw Hill Education. 2004, pp. 161–173.

[5] Bradley, P., "The History of Simulation in Medical Education and Possible Future Directions," *Medical Education*, Vol. 40, 2006, pp. 254–262.

[6] Brent, R., and Felder, R., "A Model for Engineering Faculty Development," *International Journal of Engineering Education*, Vol. 12, No. 2, 2003, pp. 234–240.

[7] Brent, R., and Felder, R., "Engineering Faculty Development: Getting the Sermon beyond the Choir," *Journal of Faculty Development*, Vol. 18, No. 3, 2001, pp. 73–81.

[8] Chin, R., and Benne, K.D., General Strategies for Effecting Changes in Human Systems. In Bennis, W.G., Benne, K.D. and Chin, R. (Eds.). *The Planning of Change* (4th edition). New York: Holt, Rinehart and Winston, 1985, pp. 32–59.

[9] Cohen, Dan., *The Heart of Change Field Guide: Tools and Tactics for Leading Change in Your Organization*. Massachusetts: Harvard Business School Press, 2005.

[10] Cooke, M., Irby, D., Sullivan, W., and Ludmerer, K., "American Medical Education 100 Years after the Flexner Report," *N Engl J Med*, Vol. 355, 2006, pp. 1339–1344.

[11] Cumming, A. and Ross, M., *The Tuning Project (Medicine) — Learning Outcomes/competences for Undergraduate Medical Education in Europe*. 2008. Edinburgh: The University of Edinburgh. Available at http://www.tuning-medicine.com/pdf/booklet.pdf (Accessed Oct 2012)

[12] Dath, D., and Iobst, W., "The Importance of Faculty Development in the Transition to Competency-based Medical Education," *Medical Teacher*, Vol. 32, No. 8, 2010, pp. 683–686.

[13] De Graaff, E. and Kolmos, A., "Characteristics of Problem-based Learning," *International Journal of Engineering Education*, Vol. 19, No. 5, 2003, pp. 657–662.

[14] Dochy, F., Segers, M., Bossche, P., Gijbels, D., "Effects of Problem-based Learning: A Meta-analysis," *Learning and Instruction*, Vol. 2003, No. 13, 2000, pp. 533–568.

[15] Dolmans, D. and Schmidt, H., "The Advantages of Problem Based Curricula," *Postgrad. Med. J.* Vol. 72, 1996, pp. 535–538.

[16] Dolmans, D., Wolfhagen, I., and Snellen-Balendong, H., "Improving the Effectiveness of Tutors in Problem-based Learning," *Medical Teacher*, Vol. 16, No. 4, 1994, pp. 369–378.

[17] Dornan, T., Carroll, C., and Parboosingh, J., "An Electronic Learning Portfolio for Reflective Continuing Professional Development," *Medical Education*, Vol. 36, No. 8, 2002, pp. 767–769.

[18] Ebert, R., and Ginzberg, E., "The Reform of Medical Education," *Health Affairs*, Vol. 7, No. 2, 1988, pp. 5–38.

[19] Enarson, C. and Burg, Frederic., "An Overview of Reform Initiatives in Medical Education: 1906 Through 1992," *JAMA*, Vol. 168, No. 9, 1992, pp. 1141–1143.

[20] Flexner, A., *Medical Education in the United States and Canada: A Report to the Carnegie Foundation for the Advancement of Teaching*. 1910. Available at http://www.carnegiefoundation.org/sites/default/files/elibrary/Carnegie_Flexner_Report.pdf (Accessed Oct 2012).

[21] Fullan, Michael. *The New Meaning of Educational Change*. New York: Teachers College Press, 2001.

[22] General Medical Council (GMC). *Tomorrow's Doctors — Outcomes and Standards for Undergraduate Medical Education.* 2009. Available at http://www.gmcuk.org/TomorrowsDoctors_ 2009.pdf_ 39260971.pdf (Accessed Oct 2012).

[23] Gibbs, G., Using Assessment strategically to Change the Way Students Learn. In S. Brown, and A. Glasner (Eds.), *Assessment Matters in Higher Education*, The Society for Research into Higher Education and Open University Press, 1999. pp. 41–53.

[24] Henriksen, Lars Bo *et al.*, *Dimension of Change: Conceptualizing Reality in Organizational Research.* Copenhagen: Copenhagen Business School Press, 2004.

[25] Institute for International Medical Education. *Global Minimum Essential Requirements in Medical Education.* Available at http://www.iime.org/documents/gmer.htm (Accessed Oct 2012).

[26] Kolmos, A. and De Graaff, E., The Process of Change to PBL. In Kolmos, A. and De Graaff, Erik (Eds.). *Management of Change.* Rotterdam: Sense Publishers. 2007, pp. 31–44.

[27] Kotter, J. B., *Why Transformation Efforts Fail.* Harvard Business Review, March–April 1995, pp. 57–67.

[28] Kotter, J., *Leading Change.* Massachusetts: Harvard Business School Press, 1996.

[29] Lam, T., and Lam, Y., "Medical Education Reform: The Asian Experience," *Academic Medicine*, Vol. 84, No. 9, 2009, pp. 1313–1317.

[30] Lam, T., Wan, X. and Ip, M. "Current Perspectives on Medical Education in China," *Medical Education*, Vol. 40, No. 10, 2006, pp. 940–949.

[31] Law, W., "The Role of the State in Higher Education Reform: Mainland China and Taiwan," *Comparative Education Review*, Vol. 39, No. 3, 1995, 322–355.

[32] Maudsley, G., "Roles and Responsibilities of the Problem Based Learning Tutor in the Undergraduate Medical Curriculum," *British Medical Journal*, Vol. 318, March 6, 1999, pp. 657–661.

[33] Maudsley, G., "Roles and Responsibilities of the Problem Based learning Tutor in the Undergraduate Medical Curriculum," *British Medical Journal*, Vol. 318, March 6, 1999, pp. 657–661.

[34] McGrath, B., Graham, I., Crotty, B. and Jolly, B., "Lack of Integration of Medical Education in Australia: The Need for Change," *Med J Aust*, Vol. 184, No. 7, 2006, pp. 346–348.

[35] Ministry of Education in China, *The Medical Education Criteria for Undergraduates*, 2005 (In Chinese).

[36] Moesby, E., "Reflections on Making a Change towards Project Oriented and Problem Based Learning," *World Transactions on Engineering and Technology Education*, Vol. 3, No. 2, 2004, pp. 269–278.

[37] Nikolova, I. and Collis, B., "Flexible learning and design of instruction," *British Journal of Educational Technology*, Vol. 29, No. 1, 1998, pp. 59–72.

[38] Oklahoma, U., "Learning Portfolios' Create Broader Awareness of Educational Achievement," *Spot Light of Teaching*, Vol. 21, No. 1, *The University of Oklahoma*, 2001.

[39] Papa, F. and Harasym, P., "Medical Curriculum Reform in North America, 1765 to the Present: A Cognitive Science Perspective," *Academic Medicine*, Vol. 74, No. 2, 1999, pp. 154–164.

[40] Paulson, F., Paulson, P., and Meyer, C., "What Makes a Portfolio a Portfolio?" *Educational Leadership*, February 1991, pp. 60–63.

[41] Qiao, M., Integrated Curriculum. In Baozhi Sun and Yuhong Zhao (Eds.). *Applied Medical Pedagogy*. Beijing: People's Medical Publishing Housing, 2011, pp. 95–102 (In Chinese).

[42] Savin-Baden, M., and Major, C., *Foundations of Problem Based Learning*. Berkshire: McGrawHill Education, 2004.

[43] Savin-Baden, M., *Facilitating Problem-based Learning: Illuminating Perspectives*. The Society for Research into Higher Education and Open University Press, Berkshire, 2003.

[44] Spencer, J., and Jordan, R., "Learner Centered Approaches in Medical Education," *BMJ*. Vol. 318(7193), May 8, 1999, pp. 1280–1283.

[45] Steinert, Y., *et al.*, "A Systematic Review of Faculty Development Initiatives Designed to Improve Teaching Effectiveness in Medical Education: BEME Guide No. 8," *Medical Teacher*, Vol. 28, No. 6, 2006, pp. 497–526.

[46] Stes, A., Clement, M., and Petegem, P.V., "The Effectiveness of A Faculty Training Programme: Long-term and Institutional Impact," *International Journal for Academic Development*, Vol. 12, No. 2, 2007, pp. 99–109.

[47] Strobel, J., and van Barneveld, A., "When is PBL More Effective? A Meta-synthesis of Meta-analyses Comparing PBL to Conventional Classrooms," *The Interdisciplinary Journal of Problem-based learning*, Vol. 3, No. 1, 2009, pp. 44–58.

[48] Sun, B., "The New Objectives of Medical Education for the 21st Century and the Reform of Higher Medical Education," *Medical Science and Philosophy*, Vol. 26, No. 2, 2005, pp. 24–26 (In Chinese).

[49] Sweeney, G., "The Challenge for Basic Science Education in Problem-based Medical Curricula," *Clin Invest Med*, Vol. 22, No. 1, 1999, pp. 15–22.

[50] Tosteson, D., "New Pathways in General Medical Education," *N Engl J Med*, Vol. 322, No. 4, 1990, pp. 234–238.

[51] Tynjälä, P., "Writing as A Tool for Constructive Learning: Students' Learning Experiences during An Experiment," *Higher Education*, Vol. 36, No. 2, 1998, pp. 209–230.

[52] Wang, G., Tai B., and Huang, C. *et al.*, "Establishing a Multidisciplinary PBL Curriculum in the School of Stomatology at Wuhan University," *International Dental Education*, Vol. 72, No. 5, 2008, pp. 610–615.

[53] Williams, G., and Lau, A., "Reform of Undergraduate Medical Teaching in the United Kingdom: A Triumph of Evangelism over Common Sense," *BMJ*, Vol. 329, 10 July 2004, pp. 92–94.

[54] Zheng, S., and Yang, M., "Discussion of Accreditation as an Opportunity to Deepen the Reform of Medical Education,"*China Higher Medical Education*, No. 7, 2010, pp. 37–38. (In Chinese).

[55] Zubizarreta, J., *The Learning Portfolio: Reflective Practice for Improving Student Learning*, Jossey-Bass Higher and Adult Education, 2004.

3

A Nation-wide Survey on Educational Reform in Medicine and Health Education in China — A Teaching Staff's Perspective

Baozhi Sun, Xiaoju Duan, Jiannong Shi, and Xiangyun Du

Abstract

This chapter gives an overview of the research methods in this study — design and contents of the survey, data generation and analysis, and limitation of the methods. This introduction is followed by descriptive data from the respondents', concerning the overall number of participants, gender, age, profession, job title, and university's geographical location. This chapter lays a methodological and empirical foundation for the following chapters of the book.

3.1 Research Methods

3.1.1 Research Background

This research project on educational reform and change in medicine and health education in China from a teaching staff's perspective was an explorative activity to prepare for the establishment of a pedagogy development program (to be established in 2012) in the North China Center for Medical Education,

Change and Reform in Medicine and Health Education in China, 39–46.

which is located in China Medical University, Shenyang, Liaoning, China. This pioneer project was aimed at reaching an understanding of teaching staff's attitudes towards the educational reforms and change in China, obtaining knowledge of the teaching staff's involvement and participation in the reforms carried out, and identifying needs and contents for planning staff/faculty development activities in China. In a broader context, this project also intends to contribute to research and activities in the relevant fields at an international level.

3.1.2 Survey Design

Supported by the North China Center for Medical Education and the China Medical Board, a survey-based investigation was conducted in late 2009. A questionnaire (see Appendix 1) on the need and current situation of medicine and health education reform was designed, amended, and finally approved by several experts in the medicine and health education field. Following the trends of both national and international standards (as introduced in Chapter 1) and inspired by the diverse innovation practices in the West, six themes were identified in the questionnaire. These themes include staff's attitudes toward and involvement in the following aspects: (1) general reforms in medicine and health education such as curricula integration, (2) pedagogy innovation such as employment of Problem-Based Learning methods, (3) new assessment methods, (4) educational technologies in the teaching practice, (5) institutional support for teaching staff's participation in educational research and reform, (6) further development of medicine and health education.

There were 58 questions in the final questionnaire. Methodologically, these 58 questions mainly fell into three types. The first type is a 5-point Likert-type scale. Participants were asked to evaluate to what extent they agreed with a statement on a scale of 1 to 5, with 1 being "totally agree" and 5 being "totally disagree". An example question is "To what extent do you think teaching and learning methods and curricula design are important parts of medicine and health education?". The data was then put into a computer and recoded, with 1 meaning "totally disagree" and 5 meaning "totally agree", this format making for data analysis easier. The second type of question was multiple choice. Participants were asked whether they agreed on the specific choices. The example question was "Which aspects of medicine and health education outcome should be further emphasized based on current indicators?". This question was provided with five options: "Practice-based learning and improvement", "Interpersonal communication skills", "Professionalism", "Diagnosis and treatment of patients", "The mastery of medical knowledge". The third type of question was simply "Yes" or "No". The example question was "Has your subject been involved in the overall curricula integration?".

3.1.3 Data Generation and Analysis

The investigation was carried out during October–November 2009. A total of 1955 participants took part in the survey and answered the questionnaire. Of these participants, 1820 people returned questionnaires which were valid (93.1%) and included in the final data analysis. The data on their gender, age, profession, job title, position, and affiliation were also collected. The software SPSS15.0 was used for data analysis. For the Likert-type questions, t-test was used for the comparison between two groups. Analysis of variance (ANOVA) was used for the comparison among three or four groups. When significant difference among these three or four groups was found, the least significant difference (LSD) post-hoc test was used. For the multiple-choice and '*Yes/No*' questions, the χ^2 test was used for analyzing. $P < .05$ was considered statistically significant. The detailed participants'distribution on gender, age, profession, job title, and university's geographical location are described in the following sections.

3.1.4 Limitation of the Methods

The design of this research is limited in a few ways for certain practical reasons. Firstly, the survey employed in the present research was a self-reporting questionnaire. The possibility of social desirability cannot be excluded. Some participants may want to be evaluated better and report biased answers to some questions. Secondly, although the 5-point Likert-type questions were one of the most popular types in scaling, participants may find it is difficult to respond using the 1-5 scale format. For the multiple-choice questions, there may be additional aspects not included in the present options. Furthermore, '*Yes/No*' questions provide very limited answers. For some questions, such as familiarity on some topics, participants have a different definition about familiarity. These "Yes" or "No" answers cannot give more information on how much they know on these topics. Thirdly, the participants of this study were recruited from normal medical universities. The process did not follow stratification sampling. The results of this study cannot be generalized to other universities, such as key universities.

3.2 Respondents' Information

3.2.1 Respondents' Backgrounds

Respondents in the investigation came from 22 medical universities in China. These universities had collaborations on different projects with the North China Center for Medical Education and the China Medical Board. Twenty of the 22 participating universities have taken part in educational reforms at different levels. The choice of participant universities (instead of choosing random universities) was based on a pragmatic approach to enable easy access to data. This has the potential limitation of shadowing the research results. However, it was also one of the aims of this study to investigate the attitudes and situation of teaching staff from the universities that have been conducting reforms. Nevertheless, the selection of teaching staff that took part in the questionnaire from each of these universities was made at random.

3.2.2 Gender

Among the 1820 valid respondent questionnaires, there were 806 males, 972 females and 42 without gender identification. Gender distribution in the study was not equally weighted. There are more females than males among the participants (see Table 3.1).

3.2.3 Age

Age was also taken into consideration in this study. One thousand and seven hundred seventeen out of the 1820 valid questionnaires stated a precise age

Table 3.1. Gender distribution.

	Gender	N	Percent	Valid Percent	χ^2	P
Valid	Male	806	44.3	45.3	15.498	<0.001
	Female	972	53.4	54.7		
	Total	1778	97.7	100.0		
Missing		42	2.3			
Total		1820	100.0			

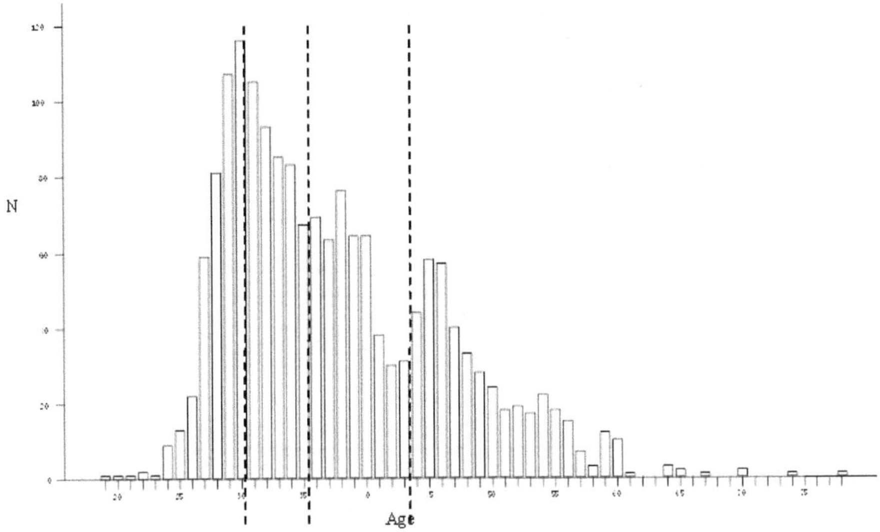

Figure 3.1. Age Distribution.

Table 3.2. Participant of each age group.

	Age group	N	Percent	Valid Percent	χ^2	P
Valid	19–30	413	22.7	24.1	0.83	0.842
	31–35	433	23.8	25.2		
	36–43	435	23.9	25.3		
	> 43	436	24.0	25.4		
	Total	1717	94.3	100.0		
Missing		103	5.7			
Total		1820	100.0			

(103 questionnaires had no age identification). In order to get similar number of participants in each age group, this sample was divided into four age ranges: ages 19–30 (413 people), ages 31–35 (433 people), ages 36–43 (435 people), and age above 43 (436 people). The participants' number across each age group is shown in the Figure 3.1.

The difference of participants' number among these four age groups was not significant (see Table 3.2).

3.2.4 Profession

These 1820 participants were from four different professions within medicine and health education, namely Teaching administrators, Basic studies teaching

Table 3.3. Participant distribution of each profession.

Profession	N	Percent	χ^2	P
Teaching administrators	128	7.0	702.26	<0.001
Basic studies teaching staff	751	41.3		
Public health teaching staff	221	12.1		
Clinical teaching staff	720	39.6		
Total	1820	100.0		

Table 3.4. Participant distribution of each job title.

Jobtitle	N	Percent	χ^2	P
Junior	438	24.1	87.430	<0.001
Middle	619	34.0		
Senior	763	41.9		
Total	1820	100.0		

staff, Public health teaching staff, and Clinical teaching staff. The groups of Basic studies teaching staff (751 people) and Clinical teaching staff (720 people) have a considerably higher number of participants than the groups of Teaching administrators (128 people) and Public health teaching staff (221 people) (see Table 3.3).

3.2.5 Job Title

Three hierarchic terminologies are often used for job titles in the Chinese academic world: Junior, Middle, and Senior in ascending order. Among the 1820 valid respondent questionnaires, all three titles were represented: 438 participants with a Junior title, 619 with a Middle title, and 763 with a Senior title (see Table 3.4). The number of each job title group was not equally distributed.

3.2.6 University Geographical Location

Participants in this investigation came from 22 medical schools/universities from different geographical areas in China (see Figure 3.2). There were significant differences concerning participants' numbers among these universities (see Table 3.5). Participants were divided in two groups according to

Figure 3.2. Participants' Universities' Geographical Distribution.

their universities' location, namely East (the eastern part of China) and West (the western part of China). This division is based on the economic situation wherein medical schools in the East generally have better financial resources than those in the West. A consequential assumption based on this is that medical schools/universities in the East have taken more advanced steps in the change process than those in the West. This assumption will be an important perspective for data analysis.

Five out of the 22 universities are located in the West, while 17 are located in the East. The number of participants from the East (1462 people) was significantly higher than in the West (359 people) (see Table 3.6).

Table 3.5. Participant distribution of each university and their location.

University No.	N	Percent	χ^2	P	Location
1	91	5.0	64.80	<0.001	East
2	90	4.9			East
3	109	6.0			East
4	79	4.3			East
5	67	3.7			East
6	76	4.2			East
7	85	4.7			East
8	85	4.7			West
9	75	4.1			East
10	77	4.2			West
11	80	4.4			East
12	111	6.1			East
13	89	4.9			East
14	50	2.7			West
15	71	3.9			East
16	76	4.2			East
17	102	5.6			East
18	60	3.3			East
19	80	4.4			East
20	68	3.7			West
21	113	6.2			West
22	86	4.7			East
Total	1820	100.0			

Table 3.6. Geographical location of participant's university.

Location	N	Percent	χ^2	P
East	1461	80.3	667.26	<0.001
West	359	19.7		
Total	1820	100.0		

3.3 Summary

This chapter introduces the background of the research methods for this book, which includes the design and contents of the survey, data generation and analysis, and the limitation of the methods. It also presents the descriptive data of the respondents' information concerning the overall number of participants, their gender, age, profession, job titles, and the universities' geographical location. This chapter functions as a methodological and empirical basis for the following chapters of the book.

4

Educational Reform — Teaching Staff's Attitudes and Involvement

Xiaoju Duan, Xiangyun Du and Jiannong Shi

Abstract

This chapter aims to gain an overall knowledge about staff's attitudes towards educational reform in medicine and health in general, and in particular, towards aspects of education that should be further emphasized and improved. It also intends to have an overview about whether the participants' subjects were involved in the curricula integration. Seven questions were asked with regard to these aspects in the survey. Findings are presented and analyzed and arranged in four sections in this chapter. Section 4.1 reports how participants responded to the importance of teaching and learning methods as well as curricula design in medicine and health education, to what extent they believed it is necessary for students in earlier study periods to access clinical courses such as Introduction to Clinical Medicine, to what extent they believed that curricula integration should involve multiple subjects, and to what extent they believed the traditional teaching method in medicine and health education should be reformed. Section 4.2 presents participants' response on which aspects of the

Change and Reform in Medicine and Health Education in China, 47–70.

current medicine and health education outcome should be further enhanced. Section 4.3 reports aspects of the current medicine and health education which should be improved in the respondents' opinions. Section 4.4 provides an overall picture of the respondents' involvement in the conducted curricula integration in medicine and health education in China. Data analyses in this chapter do not only include overall findings to the above addressed issues, but also take into consideration the impact of gender, age, profession, job title, and the universities' geographical locations, respectively.

4.1 General Opinions on Educational Reform

In order to understand staff's overall perception on the need for educational reform in medicine and health, four questions were asked that require 5-point Likert-type answers. These questions include: (1) To what extent they think teaching and learning methods and curricula design are important parts of medicine and health education (see survey Q1 Appendix 1), (2) To what extent they think it is necessary for students in earlier study periods to access clinical courses such as Introduction to Clinical Medicine (see survey Q3 Appendix 1), (3) To what extent they believe that curricula integration should involve multiple subjects (see survey Q6 Appendix 1), and (4) To what extent they believe the traditional teaching method in medicine and health education should be reformed (see survey Q7 Appendix 1).

Overall findings

Table 4.1 presents the overall findings of participants' responses to these four questions. In general, participants positively responded to all these four items about education reforms, especially to the first one (in a 5-point Likert scale with 1 to 5 points, 3 can be looked as the mean score).

The findings showed that staff who have participated in this investigation, in general, agreed on the significant importance of teaching and learning methods as well as curricula design in medical and health education in China. They also agreed on a high level of necessity for introducing Clinical Medicine in early stages of students' study, and involving multiple subjects in the curricula integration. All in all, majority of the participants suggested that the traditional teaching method in medicine and health education should be reformed.

Table 4.1. General opinions on reform.

Item	*N*	*M*	*SD*
To what extent do you think teaching and learning methods and curricula design are important parts of medicine and health education?	1805	4.60	0.53
To what extent do you think it is necessary for students in earlier study periods to access clinical courses such as Introduction to Clinical Medicine?	1664	4.09	0.80
To what extent do you believe that curricula integration should involve multiple subjects?	1798	4.22	0.71
To what extent do you believe the traditional teaching method in medicine and health education should be reformed?	1787	4.24	0.69

Influence of Gender

Both male and female groups held positive attitudes towards educational reform in medicine and health (see Table 4.2). There was no difference between male and female groups on their opinions about the importance of teaching and learning methods and curricula design, the necessity for students in earlier study periods to access clinical courses, and the need for the involvement of multiple subjects in curricula integration as well as for the reform of traditional teaching method in medicine and health education.

Table 4.2. General opinions on reform by gender.

Item	Gender	*N*	*M*	*SD*	*t*	*P*
To what extent do you think teaching and learning methods and curricula design are important parts of medicine and health education?	Male	801	4.62	0.53	1.13	0.258
	Female	962	4.59	0.52		
To what extent do you think it is necessary for students in earlier study periods to access clinical courses such as Introduction to Clinical Medicine?	Male	735	4.09	0.82	0.26	0.792
	Female	891	4.08	0.79		
To what extent do you believe that curricula integration should involve multiple subjects?	Male	799	4.21	0.74	−0.59	0.554
	Female	958	4.23	0.69		
To what extent do you believe the traditional teaching method in medicine and health education should be reformed?	Male	791	4.27	0.72	1.39	0.166
	Female	955	4.22	0.66		

Influence of Age

All the four age groups were positive to educational reforms in medicine and health education, in general. Significant differences were found within the four age groups from participants' responses to two out of the four questions: namely, their opinions on the importance of teaching and learning methods as well as curricula design, and need for involving multiple subjects in curricula integration. No significant difference was found in participants' responses to the other questions, namely, their opinions on the necessity for students in earlier study periods to access clinical courses, and the need for reforming traditional teaching method in medicine and health education. The results of variance analyses were shown in Table 4.3.

Table 4.3. General opinions on reform by age.

Item	Age group	*N*	*M*	*SD*	*F*	*P*
To what extent do you think teaching and learning methods and curricula design are important parts of medicine and health education?	19–30	412	4.71	0.48	7.18	<0.001
	31–35	433	4.61	0.51		
	36–43	428	4.59	0.55		
	> 43	431	4.55	0.55		
To what extent do you think it is necessary for students in earlier study periods to access clinical courses such as Introduction to Clinical Medicine?	19–30	381	4.18	0.80	2.10	0.099
	31–35	396	4.06	0.77		
	36–43	396	4.05	0.83		
	>43	399	4.08	0.80		
To what extent do you believe that curricula integration should involve multiple subjects?	19–30	410	4.29	0.69	3.49	0.015
	31–35	425	4.24	0.69		
	36–43	431	4.21	0.74		
	>43	431	4.14	0.72		
To what extent do you believe the traditional teaching method in medical education should be reformed?	19–30	406	4.23	0.70	0.22	0.879
	31–35	426	4.24	0.66		
	36–43	429	4.25	0.68		
	>43	425	4.27	0.72		

The *P*-values of post hoc tests between any of the two age groups are shown in Table 4.4. The post hoc analysis showed that Group 19–30 marked significantly higher degree than all the other age groups concerning the importance of learning methods and curricula design. Group19–30 and Group 31–35 marked significantly higher degree than Group above 43 concerning the need for involving multiple subjects in curricula integration.

Influence of Profession

All the four profession groups were positive towards educational changes in medicine and health. Participants from the four different profession groups reported significantly different opinions and attitudes on the first three

Table 4.4. General opinions on reform by age — post hoc analysis.

Item	19–30 vs 31–35	19–30 vs 36–43	19–30 vs >43	31–35 vs 36–43	31–35 vs >43	36–43 vs >43
To what extent do you think teaching and learning methods and curricula design are important parts of medicine and health education?	0.007	0.001	<0.001	0.514	0.070	0.248
To what extent do you think it is necessary for students in earlier study periods to access clinical courses such as Introduction to Clinical Medicine?	0.045	0.023	0.079	0.790	0.798	0.601
To what extent do you believe that curricula integration should involve multiple subjects?	0.244	0.087	0.002	0.586	0.043	0.138
To what extent do you believe the traditional teaching method in medicine and health education should be reformed?	0.867	0.671	0.444	0.795	0.544	0.728

questions, namely, the importance of teaching and learning methods and curricula design, the necessity for students in earlier study periods to access clinical courses, the involvement of multiple subjects in curricula integration. Participants of all profession groups showed no significant difference in the last question, meaning an agreement on the need for reforming traditional teaching method in medicine and health education was evidenced (see Table 4.5).

The P-values of post hoc tests between any of the two professions are shown in Table 4.6. Teaching administrators reported significantly higher degree than Basic studies teaching staff and Clinical teaching staff concerning the importance of teaching and learning methods and curricula design in medicine and health education. Concerning the necessity for students in earlier study periods to access clinical courses, Teaching administrators and Clinical teaching staff reported significantly higher degree than Basic studies teaching staff and Public health teaching staff. Basic studies teaching staff reported significantly higher than Public health teaching staff. With regards to the need for involving multiple subjects in curricula integration, Teaching administrators reported significantly higher degree than Basic studies teaching staff and Public health teaching staff. Clinical teaching staff reported significantly higher degree than Basic studies teaching staff. No significant difference

Table 4.5. General opinions on change by profession.

Item	Profession	*N*	*M*	*SD*	*F*	*P*
To what extent do you think teaching and learning methods and curricula design are important parts of medicine and health education?	Teaching administrators	127	4.72	0.45	3.64	0.012
	Basic studies teaching staff	743	4.59	0.53		
	Public health teaching staff	220	4.66	0.50		
	Clinical teaching staff	715	4.59	0.54		
To what extent do you think it is necessary for students in earlier study periods to access clinical courses such as Introduction to Clinical Medicine?	Teaching administrators	119	4.26	0.78	6.52	<0.001
	Basic studies teaching staff	676	4.06	0.81		
	Public health teaching staff	210	3.91	0.80		
	Clinical teaching staff	659	4.14	0.80		
To what extent do you believe that curricula integration should involve multiple subjects?	Teaching administrators	128	4.38	0.68	4.04	0.007
	Basic studies teaching staff	741	4.16	0.76		
	Public health teaching staff	216	4.20	0.66		
	Clinical teaching staff	713	4.25	0.68		
To what extent do you believe the traditional teaching method in medicine and health education should be reformed?	Teaching administrators	128	4.31	0.64	1.74	0.156
	Basic studies teaching staff	729	4.20	0.69		
	Public health teaching staff	219	4.25	0.68		
	Clinical teaching staff	711	4.27	0.71		

Table 4.6. General opinions on change by profession — post hoc analysis.

	P1 vs P2	P1 vs P3	P1 vs P4	P2 vs P3	P2 vs P4	P3 vs P4
To what extent do you think teaching and learning methods and curricula design are important parts of medicine and health education?	0.006	0.264	0.006	0.067	0.984	0.071
To what extent do you think it is necessary for students in earlier study periods to access clinical courses such as Introduction to Clinical Medicine?	0.010	<0.001	0.140	0.025	0.049	<0.001
To what extent do you believe that curricula integration should involve multiple subjects?	0.002	0.027	0.069	0.531	0.021	0.347
To what extent do you believe the traditional teaching method in medicine and health education should be reformed?	0.095	0.392	0.551	0.399	0.051	0.623

Note: P1, Teaching administrators; P2, Basic studies teaching staff; P3, Public health teaching staff; P4, Clinical teaching staff.

was identified between the groups of Teaching administrators and Clinical teaching staff.

Influence of Job Title

All the three job title groups were positive about these four questions. No significant differences were found within the three job title groups from participants' responses to two out of the four questions: namely, their opinions on the importance of teaching and learning methods as well as curricula design, and need for involving multiple subjects in curricula integration. However, significant differences were identified in participants responses to the other questions, namely, their opinions on the necessity for students in earlier study periods to access clinical courses, and the need for reforming traditional teaching method in medicine and health education. The findings and variance analyses are shown in Table 4.7.

The P-values of post hoc tests analyses between any of the two job title groups are shown in Table 4.8. Group with Junior job title (Junior group) reported significantly higher degree than group with Middle job title (Middle group) and group with Senior job title (Senior group) on the necessity for students in earlier study periods to access clinical courses. Senior group reported considerably higher than Middle group on need for reforming traditional teaching method in medicine and health education.

Table 4.7. General opinions on reform by job title.

Item	Job title	N	M	SD	F	P
To what extent do you think teaching and learning methods and curricula design are important parts of medicine and health education?	Junior	438	4.63	0.50	1.18	0.307
	Middle	613	4.61	0.54		
	Senior	754	4.59	0.53		
To what extent do you think it is necessary for students in earlier study periods to access clinical courses such as Introduction to Clinical Medicine?	Junior	406	4.20	0.78	6.61	0.001
	Middle	559	4.01	0.80		
	Senior	699	4.08	0.82		
To what extent do you believe that curricula integration should involve multiple subjects?	Junior	438	4.27	0.70	2.70	0.067
	Middle	607	4.24	0.71		
	Senior	753	4.17	0.72		
To what extent do you believe the traditional teaching method in medicine and health education should be reformed?	Junior	432	4.23	0.72	3.16	0.043
	Middle	607	4.20	0.68		
	Senior	748	4.29	0.69		

Table 4.8. General opinions on reform by job title–post hoc analysis.

Item	Junior vs Middle	Junior vs Senior	Middle vs Senior
To what extent do you think teaching and learning methods and curricula design are important parts of medicine and health education?	0.370	0.124	0.506
To what extent do you think it is necessary for students in earlier study periods to access clinical courses such as Introduction to Clinical Medicine?	<0.001	0.015	0.138
To what extent do you believe that curricula integration should involve multiple subjects?	0.503	0.030	0.103
To what extent do you believe the traditional teaching method in medicine and health education should be reformed?	0.503	0.130	0.014

Influence of University's Geographical Location

Participants from East (East group) and participants from West (West group) all supported educational reforms in medicine and health (see Table 4.9).

There was no significant difference between East and West group on their opinions about the importance of teaching and learning methods and curricula design, and the necessity for students in earlier study periods to access clinical

Table 4.9. General opinions on change by university's geographical location.

Item	Location	*N*	*M*	*SD*	*t*	*P*
To what extent do you think teaching and learning methods and curricula design are important parts of medicine and health education?	East	1446	4.60	0.52	−0.11	0.910
	West	359	4.61	0.53		
To what extent do you think it is necessary for students in earlier study periods to access clinical courses such as Introduction to Clinical Medicine?	East	1337	4.09	0.81	0.34	0.730
	West	327	4.07	0.78		
To what extent do you believe that curricula integration should involve multiple subjects?	East	1442	4.25	0.71	3.63	<0.001
	West	356	4.10	0.69		
To what extent do you believe the traditional teaching method in medicine and health education should be reformed?	East	1430	4.27	0.68	3.51	<0.001
	West	357	4.13	0.72		

courses. However, significant differences were found between the two groups of participants concerning their opinions on the need for involving multiple subjects in curricula integration, and the reform of traditional teaching method in medicine and health education, with the East group showing significantly higher degree than the West group.

4.2 Which Aspects of the Educational Outcome Should be Further Emphasized

In order to obtain further suggestions from the participants' concerning what concrete aspects of the education should be further emphasized and enhanced, a question with five choices was designed in the questionnaire. The five aspects were (1) Practice-based learning and improvement; (2) Interpersonal communication skills; (3) Professionalism; (4) Diagnosis and treatment of patients; and (5) The mastery of medical knowledge.

Overall findings

Participants' suggestions on which aspects of medicine and health education outcome should be further emphasized based on current indicators are shown in Table 4.10 ($N = 1813$). More than 50% participants chose Practice-based learning and improvement, Interpersonal communication skills, and Professionalism. Less than 50% participants chose Diagnosis and treatment of patients, and The mastery of medical knowledge. Based on the results, the five selections can be ranked as following: Practice-based learning and improvement, Professionalism, Interpersonal communication skills, The mastery of medical knowledge, and Diagnosis and treatment of patients.

Influence of Gender

More than 50% participants of both genders selected Practice-based learning and improvement, Interpersonal communication skills, and Professionalism. Less than 50% participants of both genders selected the choices of Diagnosis and treatment of patients, and The mastery of medical knowledge.

No significant gender difference was found as seen from the findings for all five choices with regard to what aspects of the medicine and health education should be emphasized (see Table 4.11).

Table 4.10. Which aspects of the education should be further emphasized.

Item	Yes N(%)		No N(%)	
Practice-based learning and improvement	1513	(83.5)	300	(16.5)
Interpersonal communication skills	977	(53.9)	836	(46.1)
Professionalism	1251	(69.0)	562	(31.0)
Diagnosis and treatment of patients	684	(37.7)	1129	(62.3)
The mastery of medical knowledge	859	(47.4)	954	(52.6)

Table 4.11. Which aspects of the educational outcome should be further emphasized by gender.

Item	Male N(%)		Female N(%)		χ^2	*P*
Practice-based learning and improvement	667	(82.9)	807	(83.5)	0.15	0.702
Interpersonal communication skills	445	(55.3)	508	(52.6)	1.28	0.258
Professionalism	564	(70.1)	663	(68.6)	0.42	0.516
Diagnosis and treatment of patients	323	(40.1)	345	(35.7)	3.64	0.057
The mastery of medical knowledge	389	(48.3)	450	(46.6)	0.53	0.465

The ranking of the five choices was similar for male and for female participants, which was in consistence with the overall ranking among all participants.

Influence of Age

More than 50% participants from all four age groups selected the choices of Practice-based learning and improvement, Interpersonal communication skills, and Professionalism. Less than 50% participants from all four age groups selected the choice of Diagnosis and treatment of patients. For the choice of The mastery of medical knowledge, less than 50% participants from Group 19–30, Group 31–35, and Group 36–43 selected it, while there were slightly more than 50% participants from Group above 43 who selected it.

Participants from different age groups reported similar percentages on the following aspects of medicine and health education outcomes which should be further emphasized based on current data: Practice-based learning and improvement, Interpersonal communication skills, Diagnosis and treatment of patients (see Table 4.12). Significant differences due to age factor were found on the choices of Professionalism and The mastery of medical knowledge. Findings showed an influence of age with the youngest age group agreeing less and the eldest age group agreeing most in selecting these two choices.

Table 4.12. Which aspects of the educational outcome should be further emphasized by age.

Item	19–30 N(%)		31–35 N(%)		36–43 N(%)		>43 N(%)		χ^2	*P*
Practice-based learning and improvement	342	(82.8)	367	(85.3)	366	(84.3)	349	(80.6)	3.96	0.265
Interpersonal communication skills	215	(52.1)	215	(50.0)	240	(55.3)	242	(55.9)	4.00	0.261
Professionalism	249	(60.3)	287	(66.7)	311	(71.7)	334	(77.1)	30.54	<0.001
Diagnosis and treatment of patients	135	(32.7)	155	(36.0)	168	(38.7)	174	(40.2)	5.89	0.117
The mastery of medical knowledge	183	(44.3)	186	(43.3)	200	(46.1)	228	(52.7)	9.23	0.026

The ranking of the five choices was similar in four age groups, which was consistent with the overall ranking among all participants.

Influence of Profession

For the choices of Practice-based learning and improvement, and Professionalism, more than 50% participants selected them in all these four profession groups. For the choice of Interpersonal communication skills, more than 50% participants from Teaching administrators, Public health teaching staff, Clinical teaching staff selected it, while less than 50% Basic studies teaching staff selected it. Less than 50% participants from all the profession groups selected the choice of Diagnosis and treatment of patients. Less than 50% participants from Teaching administrators, Basic studies teaching staff, and Public health teaching staff selected the choice of The mastery of medical knowledge, while slightly more than 50% Clinical teaching staff selected it.

Participants from all the four profession groups reported high choice rate on the agreement of selecting the choice of Practice-based learning and improvement as an important aspect of medicine and health education outcome (see Table 4.13). However, significant differences were found with regard to the other four choices. For the choice of Interpersonal communication skills, Teaching administrators reported highest, Clinical teaching staff and Public health teaching staff reported second and third, respectively. Basic studies teaching staff reported least. For the choice of Professionalism, Teaching administrators reported highest, Public health teaching staff and Clinical teaching staff reported second and third, respectively. Basic studies teaching staff reported least. For the choice of Diagnosis and treatment

Table 4.13. Which aspects of the educational outcome should be further emphasized by profession.

	Teaching administrators N(%)		Basic studies teaching staff N(%)		Public health teaching staff N(%)		Clinical teaching staff N(%)		χ^2	P
Practice-based learning and improvement	110	(85.9)	610	(81.7)	190	(86.0)	603	(84.1)	3.55	0.315
Interpersonal communication skills	84	(65.6)	306	(41.0)	119	(53.8)	468	(65.3)	94.70	<0.001
Professionalism	97	(75.8)	486	(65.1)	166	(75.1)	502	(70.0)	12.38	0.006
Diagnosis and treatment of patients	44	(34.4)	240	(32.1)	60	(27.1)	340	(47.4)	49.78	<0.001
The mastery of medical knowledge	49	(38.3)	336	(45.0)	92	(41.6)	382	(53.3)	18.91	<0.001

of patients, Clinical teaching staff reported highest. Teaching administrators and Basic studies teaching staff reported second and third, respectively. Public health teaching staff reported least. For the choice of The mastery of medical knowledge, Clinical teaching staff reported highest. Basic studies teaching staff reported second and Public health teaching staff reported third. Teaching administrators reported least.

A similar preference list of the five choices can be seen in all profession groups. All of the four profession groups ranked Practice-based learning and improvement as highest and Professionalism as the second highest. Interpersonal communication skills and The mastery of medical knowledge are shown as the third and fourth in the rank list of all groups except for Basic studies teaching staff. Diagnosis and treatment of patients was ranked least by all groups.

Influence of Job Title

For the choices of Practice-based learning and improvement, and Professionalism, there were more than 50% participants selecting them in all these three job title groups. For the choice of Interpersonal communication skills, there were more than 50% participants selecting it in Junior and Senior groups, while there were less than 50% Middle job title group selecting it. For the choice of Diagnosis and treatment of patients, there were less than 50% participants selecting it in all these three job title groups. For the choice of The mastery of medical knowledge, there were more than 50% participants from

Junior and Middle groups selecting it, while less than 50% participants from Senior group selecting.

Participants from the three job title groups reported an agreement on the selection of Practice-based learning and improvement (see Table 4.14). However, significant differences can be seen from their selecting of the other four choices: Interpersonal communication skills, Professionalism, Diagnosis and treatment of patients, and The mastery of medical knowledge. For the choice of Interpersonal communication skills, the Junior group ranked highest, the Senior group second, and the Middle group least. For the choice of Professionalism, the Senior group title ranked highest, the Middle group ranked second, and the Junior group ranked least.

Table 4.14. Which aspects of the educational outcome should be further emphasized by job title.

Item	Junior N(%)		Middle N(%)		Senior N(%)		χ^2	P
Practice-based learning and improvement	366	(83.6)	509	(82.8)	638	(83.9)	0.35	0.840
Interpersonal communication skills	250	(57.1)	303	(49.3)	424	(55.8)	8.18	0.017
Professionalism	276	(63.0)	411	(66.8)	564	(74.2)	18.34	<0.001
Diagnosis and treatment of patients	167	(38.1)	208	(33.8)	309	(40.7)	6.80	0.033
The mastery of medical knowledge	231	(52.7)	356	(57.9)	367	(48.3)	12.56	0.002

For the choice of Diagnosis and treatment of patients, the Senior group ranked highest, the Junior group ranked second, and the Middle group ranked least. For the choice of The mastery of medical knowledge, the Middle group ranked highest, the Junior group ranked second, and the Senior group ranked least.

A similar preference list of the five choices can be seen from the selection of all the groups. All of the three job title groups ranked Practice-based learning and improvement as highest and Professionalism as the second highest. Interpersonal communication skills were ranked as the third by both Junior and Senior groups. The mastery of medical knowledge was chosen as the third by the Middle group. Diagnosis and treatment of patients was ranked least by all groups.

Influence of University's Geographical Location

For the choices of Practice-based learning and improvement, Interpersonal communication skills, and Professionalism, there were more than 50%

participants selecting them in both East and West groups. For the choice of Diagnosis and treatment of patients, there were less than 50% participants selecting it in both East and West groups. For the choice of The mastery of medical knowledge, less than 50% participants from East group chose it, while there were slightly more than 50% participants from West group selecting it.

With regard to university's geographical location, no significant differences were identified between East and West groups in three out of the five choices: Practice-based learning and improvement, Diagnosis and treatment of patients, and The mastery of medical knowledge (see Table 4.15). Participants from the two groups reported significant difference in the two choices of Interpersonal communication skills and Professionalism, with the West group reporting higher than the East group.

Table 4.15. Which aspects of the educational outcome should be further emphasized by university's geographical location.

Item	East N(%)		West N(%)		χ^2	*P*
Practice-based learning and improvement	1215	(83.4)	298	(83.5)	0.00	0.991
Interpersonal communication skills	768	(52.7)	209	(58.5)	3.88	0.049
Professionalism	982	(67.4)	269	(75.4)	8.38	0.004
Diagnosis and treatment of patients	544	(37.4)	140	(39.2)	0.42	0.517
The mastery of medical knowledge	675	(46.4)	184	(51.5)	3.09	0.079

The findings also showed a similar preference list of the five choices in two geographical groups: Practice-based learning and improvement as the highest, Professionalism as the second highest, Interpersonal communication skills as the third highest, The mastery of medical knowledge and Diagnosis and treatment of patients as the lowest.

4.3 Which Aspects of the Education Should be Improved

In order to obtain further suggestions from the participants' concerning what aspects of should be improved in medicine and health education, a multiple-choice question was asked with four choices to select. The choices were: (1) The connection between different parts of knowledge, (2) Curricula integration, (3) Emphasis on both teaching and learning, and (4) Knowledge memorization.

Overall findings

Participants' options on which aspects should be improved in medicine and health education based on the current situation are shown in Table 4.16 ($N = 1805$). There were more than 50% participants selecting choices of The connection between different parts of knowledge, Curricula integration, Emphasis on both teaching and learning. There were less than 50% participants selecting the choice of Knowledge memorization. Results of participants selection in the four choices can be ranked as following: The connection between different parts of knowledge, Curricula integration, Emphasis on both teaching and learning, Knowledge memorization.

Table 4.16. Which aspects of the education should be improved.

Item	Yes N(%)		No N(%)	
The connection between different parts of knowledge	1332	(73.8)	473	(26.2)
Curricula integration	1119	(62.0)	686	(38.0)
Emphasis on both teaching and learning	957	(53.0)	848	(47.0)
Knowledge memorization	366	(20.3)	1439	(79.7)

Influence of Gender

Generally speaking, there were more than 50% participants selecting choices of The connection between different parts of knowledge, Curricula integration, Emphasis on both teaching and learning in both male and female groups. There were less than 50% participants selecting the choice of Knowledge memorization in both male and female groups.

There was significant difference between male and female groups on the choice of The connection between different parts of knowledge. Female group reported higher percentage than male group. Male and female groups reported similar percentage on the other three choices: Curricula integration, Emphasis on both teaching and learning, Knowledge memorization (see Table 4.17).

Table 4.17. Which aspects of the education should be improved by gender.

Item	Male N(%)		Female N(%)		χ^2	P
The connection between different parts of knowledge	570	(71.4)	734	(76.1)	4.87	0.027
Curricula integration	503	(63.0)	593	(61.5)	0.47	0.495
Emphasis on both teaching and learning	434	(54.4)	500	(51.8)	1.16	0.281
Knowledge memorization	178	(22.3)	182	(18.9)	3.19	0.074

The ranking of four items was similar for male and female group, which match the overall ranking of all participants.

Influence of Age

More than 50% participants from all four age groups selected the choices of The connection between different parts of knowledge, Curricula integration. The choice of Emphasis on both teaching and learning was selected by less than 50% participants from Groups of 19–30, and 31–35, but selected by more than 50% participants from Groups of 36–43, and above 43. The choice of Knowledge memorization was selected by less than 50% participants from all four age groups.

Participants in different age groups reported similar percentages on the following aspects that should be improved in medicine and health education based on the current situation: The connection between different parts of knowledge, and Curricula integration (see Table 4.18). The result of the choice of Emphasis on both teaching and learning showed as a function of age with the youngest age group agreeing least and the eldest age group agreeing most. For the choice of Knowledge memorization, Group above 43 reported highest. Group 36–43 and Group 19–30 reported second and third, respectively. Group 31–35 reported least.

The ranking of four choices was similar in all these four age groups, which was consistent with the overall ranking of all participants.

Influence of Profession

For the choices of The connection between different parts of knowledge, Curricula integration, there were more than 50% participants selecting these

Table 4.18. Which aspects of the education should be improved by age.

Item	19–30 N(%)		31–35 N(%)		36–43 N(%)		>43 N(%)		χ^2	P
The connection between different parts of knowledge	293	(71.3)	323	(75.3)	325	(75.6)	312	(72.2)	3.05	0.384
Curricula integration	264	(64.2)	262	(61.1)	271	(63)	259	(60)	1.99	0.575
Emphasis on both teaching and learning	193	(47.0)	208	(48.5)	231	(53.7)	265	(61.3)	21.62	<0.001
Knowledge memorization	77	(18.7)	68	(15.9)	89	(20.7)	105	(24.3)	10.19	0.017

choices in all four profession groups. For the choice of Emphasis on both teaching and learning, there were more than 50% participants from groups of Teaching administrators, Public health teaching staff, and Clinical teaching staff selecting it, while there were less than 50% participants from group of Basic studies teaching staff selecting it. For the choice of Knowledge memorization, less than 50% participants from all four profession groups selected it.

Participants with different professions reported different percentages on the following choices: The connection between different parts of knowledge, Curricula integration, and Knowledge memorization. For the choice of The connection between different parts of knowledge, Clinical teaching staff reported highest. Basic studies teaching staff and Public health teaching staff were second and third, respectively. Teaching administrators reported least. For the choice of Curricula integration, Teaching administrators reported highest, while Public health teaching staff and Clinical teaching staff reported second and third, respectively. Basic studies teaching staff remained the least. For the choice of Knowledge memorization, Clinical teaching staff reported highest. Basic studies teaching staff and Teaching administrators reported second and third, respectively, while Public health teaching staff remained the least. Participants' opinions on the choice of Emphasis on both teaching and learning were similar among these four professions (see Table 4.19).

A similar preference list of the four choices can be seen from the selections of the four profession groups. Teaching administrators and Public health teaching staff ranked Curricula integration as highest and The connection between different parts of knowledge as the second highest. Basic studies teaching staff and Clinical teaching staff ranked The connection between different parts of

Table 4.19. Which aspects of the education should be improved by profession.

	Teaching administrators N(%)		Basic studies teaching staff N(%)		Public health teaching staff N(%)		Clinical teaching staff N(%)		χ^2	P
The connection between different parts of knowledge	87	(68.0)	541	(72.8)	150	(68.2)	554	(77.6)	11.52	0.009
Curricula integration	98	(76.6)	416	(56.0)	152	(69.1)	453	(63.4)	28.24	<0.001
Emphasis on both teaching and learning	73	(57.0)	368	(49.5)	122	(55.5)	394	(55.2)	6.33	0.097
Knowledge memorization	19	(14.8)	114	(15.3)	32	(14.5)	201	(28.2)	45.38	<0.001

knowledge as the highest and Curricula integration as the second highest. All these four profession groups ranked Emphasis on both teaching and learning as the second lowest and Knowledge memorization as the lowest.

Influence of Job Title

For the choices of The connection between different parts of knowledge, Curricula integration, there were more than 50% participants selecting them in all these three job title groups. For the choice of Emphasis on both teaching and learning, there were more than 50% participants from Junior and Senior groups, while there were less than 50% participants from Middle group selecting it. For the choice of Knowledge memorization, less than 50% participants from all three job title groups selected it.

Participants with different job titles reported similar percentages on choices of The connection between different parts of knowledge, and Curricula integration. Participants' opinions on the choices of Emphasis on both teaching and learning, and Knowledge memorization were different among these three job titles, with Senior group reported highest, Junior group reported second and the Middle group remained the lowest (see Table 4.20).

A same preference list of the four choices can be seen from the selections of all the three job title groups. They ranked The connection between different parts of knowledge as highest and Curricula integration as the second highest. Emphasis on both teaching and learning was ranked as the third and Knowledge memorization as the lowest by all groups.

Influence of University's Geographical Location

Generally speaking, more than 50% participants from both East and West groups selected the choices of the connection between different parts

Table 4.20. Which aspects of the education should be improved by job title.

Item	Junior N(%)		Middle N(%)		Senior N(%)		χ^2	*P*
The connection between different parts of knowledge	319	(73.2)	444	(72.2)	569	(75.5)	1.99	0.370
Curricula integration	285	(65.4)	371	(60.3)	463	(61.4)	2.94	0.230
Emphasis on both teaching and learning	227	(52.1)	290	(47.2)	440	(58.4)	17.27	<0.001
Knowledge memorization	98	(22.5)	91	(14.8)	177	(23.5)	17.50	<0.001

of knowledge, Curricula integration, and Emphasis on both teaching and learning. There were less than 50% participants selecting the choice of Knowledge memorization in both East and West groups.

With regards to university's geographical location, no significant differences were found between East and West groups in three out of the four choices: The connection between different parts of knowledge, Curricula integration, and Knowledge memorization (see Table 4.21). Participants from the two groups reported significant differences in the choice of Emphasis on both teaching and learning with West group reported higher than East group did.

Table 4.21. Which aspects of the education should be improved by university's geographical location.

Item	East N(%)		West N(%)		χ^2	P
The connection between different parts of knowledge	1058	(73.0)	274	(77.2)	2.62	0.105
Curricula integration	897	(61.9)	222	(62.5)	0.06	0.815
Emphasis on both teaching and learning	743	(51.2)	214	(60.3)	9.36	0.002
Knowledge memorization	283	(19.5)	83	(23.4)	2.63	0.105

The same preference list of the four choices can be seen from the two groups. They ranked The connection between different parts of knowledge as highest and Curricula integration as the second highest. Emphasis on both teaching and learning was ranked third and Knowledge memorization as the lowest by all groups.

4.4 Current Inplementation of Curricula Integration

This section gives an overview of the participants' involvement in the overall curricula integration that has been conducted in medicine and health education in China.

Overall findings

One question was asked (see survey Q5 Appendix 1) and findings are shown in Table 4.22. Less than 50% participants chose '*Yes*' answer indicating that

Table 4.22. Whether one's subject is involved in the overall curricula integration.

Item	N	Yes N(%)	No N(%)
Has your subject been involved in the overall curricula integration?	1755	726 (41.4)	1029 (58.6)

the majority of participants' subject was not involved in the overall curricula integration.

Influence of Gender

Less than 50% participants chose '*Yes*' answer in both male and female groups. No significant gender difference was shown from the findings (see Table 4.23).

Table 4.23. Whether one's subject is involved in the overall curricula integration by gender.

Item	Male N(%)		Female N(%)		χ^2	*P*
Has your subject been involved in the overall curricula integration?	338	(43.3)	373	(39.9)	2.02	0.155

Note: The percentages presented in this table are percentage of option 'Yes' and same in the following tables unless it is indicated.

Influence of Age

In general, there were less than 50% participants choosing '*Yes*' answer in all these four age groups. No significant difference due to age was shown from the findings (see Table 4.24).

Table 4.24. Whether one's subject is involved in the overall curricula integration by age.

Item	19–30 N(%)		31–35 N(%)		36–43 N(%)		>43 N(%)		χ^2	*P*
Has your subject been involved in the overall curricula integration?	166	(41.7)	177	(41.9)	177	(42.3)	173	(41.4)	0.08	0.994

Influence of Profession

In general, there were less than 50% participants choosing '*Yes*' answer in all these four profession groups. Significant differences among profession groups were shown from the findings (see Table 4.25).

Table 4.25. Whether one's subject is involved in the overall curricula integration by profession.

	Teaching administrators N(%)		Basic studies teaching staff N(%)		Public health teaching staff N(%)		Clinical teaching staff N(%)		χ^2	*P*
Has your subject been involved in the overall curricula integration?	47	(38.5)	313	(43.4)	59	(28.1)	307	(43.7)	18.52	< 0.001

A significantly higher percentage of participants from Basic studies teaching staff and Clinical teaching staff than from Teaching administrators and Public health teaching staff reported that their subjects were involved in the overall curricula integration.

Influence of Job Title

In general, there were less than 50% participants choosing '*Yes*' answer in all these three job title groups. No significant difference due to job titles was shown from the findings (see Table 4.26).

Table 4.26. Whether one's subject is involved in the overall curricula integration by job title.

Item	Junior N(%)		Middle N(%)		Senior N(%)		χ^2	P
Has your subject been involved in the overall curricula integration?	192	(45.0)	235	(39.8)	299	(40.6)	3.10	0.212

Influence of University's Geographical Location

In general, there were less than 50% participants choosing '*Yes*' answer in both East and West groups. With regard to university's geographical location, significant difference was shown between East and West groups, with the East group higher than the West (see Table 4.27).

Table 4.27. Whether one's subject is involved in the overall curricula integration by university's geographical location.

Item	East N(%)		West N(%)		χ^2	P
Has your subject been involved in the overall curricula integration?	619	(44.1)	107	(30.5)	21.43	< 0.001

4.5 Summary

In order to have an overall knowledge about staff's attitudes towards education reform in medicine and health, this chapter investigated four aspects: first, the participants' overall response to reform: to what extent they believed in the importance of teaching and learning methods as well as curricula design in medicine and health education, to what extent they believed it is necessary for students in earlier study periods to access clinical courses such as Introduction to Clinical Medicine, to what extent they believed that curricula integration should involve multiple subjects, and to what extent they believed the traditional teaching method in medicine and health education should be reformed. Second, which aspects of the current medicine and health educational outcome should be further emphasized. Third, which aspects of education should be improved. Fourth, whether the participants' subjects were involved in the conducted curricula integration in China.

Seven questions were asked with regard to these aspects in the survey, which included four 5-point Likert-type questions, two multiple-choice questions and one *'Yes/No'* question. Overall findings to these questions were reported in the four sections in this chapter, and variables such as gender, age, profession, job title, and the university's geographical location were respectively taken into consideration of analysis.

Overall findings in this chapter can be summarized in the following:

1. Staff who have participated in this investigation in general agreed on the significant importance of teaching and learning methods as well as curricula design in medical and health education in China. Agreements on a high level of necessity for introducing Clinical Medicine in the early stages of students' study, and involving multiple subjects in the curricula integration were clearly observed among the majority of the participants. All in all, majority of the participants suggested that the traditional teaching method in medicine and health education should be reformed. The positive general agreement on these aspects are despite the factors of gender, age, profession, job title and geographical location of their universities. These findings show a positive overall picture of staff's support for educational reform in medicine and health. It seems that there is a 'U' shaped tendency of opinions on educational reform in medicine and health education on participants of different job titles. In other words, junior and senior staff are more supportive for educational reform than staff with middle job titles.
2. Two out of the five listed educational outcomes in medicine and health were agreed by more than 50% of the respondents, which are Practice-based learning and improvement (83.5%) and Professionalism (69%). Less than half of the participants believed

that Interpersonal communication skills, Mastery of medical knowledge and Diagnosis and treatment of patients should be further emphasized as important medicine and health educational outcomes. This indicates a general agreement of staff's support for strengthening new skills and competences as the expected educational outcome, which is consistent with the trends of change in the world (see Chapter 2). Both the positive and negative agreements in the above mentioned aspects were despite the factors of gender, age, job title and geographical location of their universities. Nevertheless, except for the groups of teaching administrators and clinical teaching staff, Interpersonal communication skills, which is also an important learning outcome suggested in the Tomorrow's Doctor, Tuning project, among others in the western world, is not so far acknowledged by the staff in China. These findings also suggest a general trend that the younger the participants, the more positive they are toward educational reform.

3. Concerning which aspects can be further improved in medicine and health education, two out of the four listed choices were positively agreed by more than half of the respondents: The connection between different parts of knowledge (73.8%) and Curricula integration (62%). This positive agreement was despite gender, age, profession, job title and geographical location of their universities, except for Basic studies teaching staff's negative response to Curricula integration. Less than half of the participants responded positively to Emphasis on both teaching and learning (53%). 20.3% of participants from all groups reported positively about Knowledge memorization. These findings indicate a general agreement on the two concrete aspects of educational reform. Nevertheless, the limited support for emphasis on both teaching and learning showed a lack of focus on students' learning in general.
4. Less than half of the participants (41.4%) participated in the overall curricula integration that had been conducted in medicine and health education in China. None of the groups in relation to the factors of gender, age, profession, job title and geographical location of their universities reported more than 50% participation in the conducted curricula integration.

To conclude, a strong agreement was evidenced that there is a need for reform and change, in general, in medicine and health education in China. Nevertheless, factors such as age, profession, job title and geographical location of their universities made a difference in the participants' attitudes towards concrete aspects of change.

5

Pedagogy Innovation — Teaching Staff's Attitudes and Participation

Xiaoju Duan, Xiangyun Du and Jiannong Shi

Abstract

This chapter reports teaching staff's attitudes towards and participation in pedagogy innovation. It aims to get overall ideas of staff concerning introducing new or alternative teaching methods to the current medicine and health education, and about the familiarity of particular teaching methods. Eleven questions were asked regarding these aspects. Data are analyzed with SPSS 15.0 and the results are presented in this chapter, which are structured in two sections. Section 1 reports how participants respond to (1) to what extent they think alternative teaching methods (e.g. PBL) should be introduced to medicine and health education; (2) to what they think of the effect of PBL; (3) to what extent they agree that reflective learning methods should be implemented in medical teaching; and (4) to what extent they think that the concept of lifelong learning should be integrated in the teaching and learning process. Section 2 presents the involvement and familiarity of participants' universities in the following aspects: (1) whether the PBL has been implemented at their university; (2) whether they are familiar with the concepts of mentoring and student support;

Change and Reform in Medicine and Health Education in China, 71–86.

(3) whether they are familiar with the concepts of reflection and reflective practice; (4) whether they use reflective learning methods in their teaching; (5) whether they are familiar with the concept of lifelong learning; (6) whether they are familiar with the concept of co-operative learning; and (7) whether they are familiar with the concept of portfolio building. The overall findings, as well as the effects of participants' gender, age, profession, job title, and their university's location, on the above mentioned issues are presented, respectively.

5.1 General Opinions on Teaching Methods

In order to understand teaching staff's overall perception on the teaching methods in medicine and health education, four questions were asked that require 5-point Likert-type answers. These questions include: (1) To what extent they think alternative teaching methods (e.g. PBL) should be introduced to medicine and health education (see survey Q8 Appendix 1); (2) What they think of the effect of PBL (see survey Q10 Appendix 1); (3) To what extent they agree that reflective learning methods should be implemented in medical teaching (see survey Q14 Appendix 1); and (4) To what extent they think that the concept of lifelong learning should be integrated in the teaching and learning process (see survey Q16 Appendix 1).

Overall findings

The overall descriptive data of participants' opinions on the introduction of alternative teaching methods (e.g. PBL) to medicine and health education, the effect of PBL, the implementation of reflective learning methods in medical teaching, the integration of lifelong learning in the teaching and learning process, is shown in Table 5.1. In general, participants positively responded on all these four questions, especially on the last one. The results indicated that the participants held a positive attitude towards new teaching methods in medicine and health education, such as PBL, reflective learning, and lifelong learning.

Influence of Gender

Both male and female participants were positive to new teaching methods in medicine and health education (see Table 5.2). They had similar opinions on the introduction of alternative teaching methods in medicine and health

Table 5.1. General opinions on teaching method.

	N	*M*	*SD*
To what extent do you think alternative teaching methods (e.g. PBL) should be introduced to medicine and health education?	1804	4.27	0.69
What do you think of the effect of PBL?	1732	4.04	0.70
To what extent do you agree that reflective learning methods should be implemented in medical teaching?	1782	4.14	0.71
To what extent do you think that the concept of lifelong learning should be integrated in the teaching and learning process?	1805	4.51	0.64

Table 5.2. General opinions on teaching method by gender.

Item	Gender	*N*	*M*	*SD*	*t*	*P*
To what extent do you think alternative teaching methods (e.g. PBL) should be introduced to medicine and health education?	Male	803	4.27	0.69	−0.01	0.989
	Female	959	4.27	0.69		
What do you think of the effect of PBL?	Male	774	4.03	0.69	−0.27	0.785
	Female	919	4.04	0.70		
To what extent do you agree that reflective learning methods should be implemented in medical teaching?	Male	795	4.16	0.73	1.25	0.210
	Female	947	4.12	0.71		
To what extent do you think that the concept of lifelong learning should be integrated in the teaching and learning process?	Male	799	4.52	0.64	0.61	0.545
	Female	965	4.50	0.63		

education, PBL's effect, the implementation of reflective teaching method, and the integration of lifelong learning in the teaching and learning process. No gender differences were found on these aspects.

Influence of Age

All the four age groups held positive attitudes to new teaching methods in medicine and health education. Participants of different age groups had similar opinions on the introduction of alternative teaching methods in medicine and health education, PBL's effect, the implementation of reflective teaching method, and the integration of lifelong learning in the teaching and learning process (see Table 5.3).

In order to see if there is any difference between age groups, the post hoc tests between any two age groups were conducted and the *P*-values are presented in Table 5.4.

Table 5.3. General opinions on teaching method by age.

Item	Age group	*N*	*M*	*SD*	*F*	*P*
To what extent do you think alternative teaching methods (e.g. PBL) should be introduced to medicine and health education?	19–30	410	4.22	0.77	1.41	0.239
	31–35	428	4.30	0.65		
	36–43	433	4.30	0.66		
	>43	432	4.25	0.70		
What do you think of the effect of PBL?	19–30	395	4.02	0.71	1.48	0.219
	31–35	412	4.08	0.67		
	36–43	417	3.99	0.76		
	>43	412	4.07	0.64		
To what extent do you agree that reflective learning methods should be implemented in medical teaching?	19–30	404	4.16	0.70	0.28	0.839
	31–35	425	4.14	0.72		
	36–43	430	4.14	0.71		
	>43	422	4.11	0.72		
To what extent do you think that the concept of lifelong learning should be integrated in the teaching and learning process?	19–30	410	4.50	0.65	1.84	0.137
	31–35	431	4.50	0.64		
	36–43	432	4.47	0.67		
	>43	431	4.57	0.59		

Table 5.4. General opinions on teaching method by age-post hoc analysis.

Item	19–30 vs 31–35	19–30 vs 36–43	19–30 vs >43	31–35 vs 36–43	31–35 vs vs >43	36–43 vs >43
To what extent do you think alternative teaching methods (e.g. PBL) should be introduced to medicine and health education?	0.089	0.103	0.592	0.941	0.238	0.267
What do you think of the effect of PBL?	0.241	0.543	0.305	0.071	0.881	0.098
To what extent do you agree that reflective learning methods should be implemented in medical teaching?	0.731	0.777	0.371	0.951	0.577	0.535
To what extent do you think that the concept of lifelong learning should be integrated in the teaching and learning process?	0.891	0.531	0.107	0.440	0.135	0.023

Influence of Profession

In order to explore if there is any difference between participants with different professions, data were analyzed based on four professions, say Teaching administrators, Basic studies teaching staff, Public health teaching staff, and Clinical teaching staff. The result showed that, in general, all four profession groups very positively responded to new teaching methods in medicine and health education. There were significant profession effects on the first three questions, namely, the introduction of alternative teaching methods in medicine and health education, PBL's effect, the implementation of reflective teaching method. No profession effect was found on the last question about the integration of lifelong learning in the teaching and learning process (see Table 5.5).

In order to detect where the profession effects came from, the post hoc tests between any two profession groups were conducted and the P-values are shown in Table 5.6. Teaching administrators evaluated higher on the introduction of alternative teaching methods in medicine and health education than the other three profession groups. Concerning the PBL's effect, Teaching administrators and Clinical teaching staff reported higher than Basic studies teaching staff. With regard to the implementation of reflective learning method, Teaching administrators got higher score than Public health teaching staff, Clinical teaching staff got higher score than both Basic studies teaching staff and Public health teaching staff. On the item about integration of lifelong learning

Table 5.5. General opinions on teaching method by profession.

Item	Profession	*N*	*M*	*SD*	*F*	*P*
To what extent do you think alternative teaching methods (e.g. PBL) should be introduced to medicine and health education?	Teaching administrators	126	4.43	0.66	3.54	0.014
	Basic studies teaching staff	742	4.22	0.71		
	Public health teaching staff	220	4.28	0.75		
	Clinical teaching staff	716	4.29	0.66		
What do you think of the effect of PBL?	Teaching administrators	124	4.13	0.64	4.72	0.003
	Basic studies teaching staff	713	3.96	0.71		
	Public health teaching staff	214	4.05	0.59		
	Clinical teaching staff	681	4.09	0.71		
To what extent do you agree that reflective learning methods should be implemented in medical teaching?	Teaching administrators	124	4.20	0.72	5.41	0.001
	Basic studies teaching staff	729	4.10	0.71		
	Public health teaching staff	217	4.02	0.72		
	Clinical teaching staff	712	4.21	0.71		
To what extent do you think that the concept of lifelong learning should be integrated in the teaching and learning process?	Teaching administrators	128	4.66	0.55	2.57	0.053
	Basic studies teaching staff	744	4.50	0.63		
	Public health teaching staff	219	4.48	0.61		
	Clinical teaching staff	714	4.50	0.66		

Table 5.6. General opinions on teaching method by profession — post hoc analysis.

Item	P1 vs P2	P1 vs P3	P1 vs P4	P2 vs P3	P2 vs P4	P3 vs P4
To what extent do you think alternative teaching methods (e.g. PBL) should be introduced to medicine and health education?	0.002	0.050	0.033	0.300	0.077	0.865
What do you think of the effect of PBL?	0.014	0.293	0.560	0.124	0.001	0.430
To what extent do you agree that reflective learning methods should be implemented in medical teaching?	0.154	0.022	0.896	0.125	0.004	0.001
To what extent do you think that the concept of lifelong learning should be integrated in the teaching and learning process?	0.011	0.012	0.009	0.654	0.901	0.718

Note: P1, Teaching administrators; P2, Basic studies teaching staff; P3, Public health teaching staff, P4, Clinical teaching staff.

into teaching and learning process, the group of Teaching administrators got higher score than the other three groups and there are no statistical significant differences among the other three groups.

Influence of Job Title

All the three job title groups were positive about these four questions. Participants with different job titles had similar opinions on three out of the four questions: namely, the introduction of alternative teaching methods in medicine and health education, the implementation of reflective teaching method, and integration of lifelong learning in the teaching and learning process. Participants with different job titles had different opinions on PBL's effect (see Table 5.7).

The P-values of post hoc tests between any two title groups are shown in Table 5.8. Junior and Senior groups responded more positively on PBL's effect than Middle group.

Influence of University's Geographical Location

In general, both East and West groups had positive opinions on new teaching methods introduced in medicine and health education. But, there were significant geographical location effects on each item. The results listed in Table 5.9

Table 5.7. General opinions on teaching method by job title.

Item	Job title	*N*	*M*	*SD*	*F*	*P*
To what extent do you think alternative teaching methods (e.g. PBL) should be introduced to medicine and health education?	Junior	435	4.24	0.75	0.62	0.540
	Middle	612	4.28	0.67		
	Senior	757	4.28	0.67		
What do you think of the effect of PBL?	Junior	415	4.07	0.74	4.90	0.008
	Middle	593	3.96	0.72		
	Senior	724	4.07	0.65		
To what extent do you agree that reflective learning methods should be implemented in medical teaching?	Junior	432	4.20	0.70	1.80	0.166
	Middle	603	4.12	0.71		
	Senior	747	4.13	0.72		
To what extent do you think that the concept of lifelong learning should be integrated in the teaching and learning process?	Junior	435	4.49	0.64	0.74	0.475
	Middle	616	4.50	0.66		
	Senior	754	4.53	0.62		

Table 5.8. General opinions on teaching method by job title-post hoc analysis.

Item	Junior vs Middle	Junior vs Senior	Middle vs Senior
To what extent do you think alternative teaching methods (e.g. PBL) should be introduced to medicine and health education?	0.325	0.313	0.986
What do you think of the effect of PBL?	0.014	0.983	0.004
To what extent do you agree that reflective learning methods should be implemented in medical teaching?	0.083	0.096	0.876
To what extent do you think that the concept of lifelong learning should be integrated in the teaching and learning process?	0.813	0.275	0.348

Table 5.9. General opinions on teaching method by location.

Item	Location	*N*	*M*	*SD*	*t*	*P*
To what extent do you think alternative teaching methods (e.g. PBL) should be introduced to medicine and health education?	East	1448	4.28	0.69	1.95	0.052
	West	356	4.21	0.68		
What do you think of the effect of PBL?	East	1399	4.06	0.71	2.70	0.007
	West	333	3.94	0.64		
To what extent do you agree that reflective learning methods should be implemented in medical teaching?	East	1429	4.17	0.71	2.94	0.003
	West	353	4.04	0.74		
To what extent do you think that the concept of lifelong learning should be integrated in the teaching and learning process?	East	1450	4.54	0.61	3.68	<0.001
	West	355	4.40	0.73		

shows that participants from East universities got higher scores than participants from West universities on alternative teaching methods (eg. PBL) that should be introduced into medicine and health education (marginal significant), on effects of PBL, on the implementation of reflective teaching method, and on integration of lifelong learning in the teaching and learning process (see Table 5.9).

5.2 Staff's Knowledge About New Teaching Methods

In order to understand staff's knowledge about new teaching methods in medicine and health education, seven '*Yes/No*' questions were asked. These questions include: (1) Has PBL been implemented at their university (see survey Q9 Appendix 1)? (2) Are they familiar with the concepts of mentoring and student support (see survey Q11 Appendix 1)? (3) Are they familiar with the concepts of reflection and reflective practice (see survey Q12 Appendix 1)? (4) Do they use reflective learning methods in their teaching (see survey Q13 Appendix 1)? (5) Are they familiar with the concept of lifelong learning (see survey Q15 Appendix 1)? (6) Are they familiar with the concept of co-operative learning (see survey Q17 Appendix 1)? (7) Are they familiar with the concept of portfolio building (see survey Q18 Appendix 1)?

Overall findings

The descriptive data of participants' options on the implementation of PBL in their universities, the familiarity of mentoring and student support, the familiarity of reflection and reflective practice, the use of reflective learning method in their own teaching, the familiarity of lifelong learning, and the familiarity of co-operative learning, the familiarity of portfolio building, is shown in Table 5.10. In general, there were more than 50% participants answering '*Yes*' on five out of seven questions: namely, the implementation of PBL in their universities, the familiarity of lifelong learning, and the familiarity of co-operative learning. There were less than 50% participants answering '*Yes*' on two out of the seven questions: namely, the familiarity of mentoring and student support, the familiarity of portfolio building.

Table 5.10. Staff's knowledge about new teaching methods.

Item	*N*	Yes	No
Has PBL been implemented at your university?	1785	1351 (75.7%)	434 (24.3%)
Are you familiar with the concepts of mentoring and student support?	1796	827 (46.0%)	969 (54.0%)
Are you familiar with the concepts of reflection and reflective practice?	1810	936 (51.7%)	874 (48.3%)
Do you use reflective learning methods in your teaching?	1777	907 (51.0%)	870 (49.0%)
Are you familiar with the concept of lifelong learning?	1798	1592 (88.5%)	206 (11.5%)
Are you familiar with the concept of co-operative learning?	1806	1252 (69.3%)	554 (30.7%)
Are you familiar with the concept of portfolio building?	1804	727 (40.3%)	1077 (59.7%)

Influence of Gender

For the implementation of PBL in their universities, there were more than 70% participants choosing '*Yes*' answer in both male and female group. Concerning the familiarity of mentoring and student support, the male group reported 50.4%, while female group reported 42.8% on '*Yes*' option, and there was significant difference between male and female groups. With regard to the familiarity of reflection and reflective practice, male group reported (54.9%) significantly higher than female group did (49.1%) on '*Yes*' option. As to the use of reflective learning methods, there was no significant difference between male (53.3%) and female groups (48.8%). For the familiarity of lifelong learning, there were more than 80% participants choosing '*Yes*' answer in both male and female groups. For the familiarity of co-operative learning, there were more than 60% participants choosing '*Yes*' answer in both male and female groups. There were less than 50% participants choosing '*Yes*' answer in both male and female groups on the familiarity of portfolio building. Male group reported higher percentage than female group on this question (see Table 5.11).

Influence of Age

For the implementation of PBL in their universities, there were more than 70% participants choosing '*Yes*' answer in all the four age groups. The difference among these four age groups was not significant. Concerning the familiarity of mentoring and student support, there was significant age effect. Group above

Table 5.11. Staff's knowledge about new teaching methods by gender.

Item	Male N(%)		Female N(%)		χ^2	*P*
Has PBL been implemented at your university?	586	(74.2)	729	(76.5)	1.25	0.263
Are you familiar with the concepts of mentoring and student support?	400	(50.4)	411	(42.8)	10.01	0.002
Are you familiar with the concepts of reflection and reflective practice?	440	(54.9)	475	(49.1)	5.92	0.015
Do you use reflective learning methods in your teaching?	422	(53.3)	461	(48.8)	3.41	0.065
Are you familiar with the concept of lifelong learning?	703	(88.1)	854	(89.1)	0.40	0.530
Are you familiar with the concept of co-operative learning?	554	(69.0)	669	(69.5)	0.05	0.828
Are you familiar with the concept of portfolio building?	356	(44.5)	354	(36.8)	10.88	0.001

Note: The percentages presented in this table are percentage of option 'Yes' and same in the following tables unless it is indicated.

43 reported highest (53.1%), while Group 19–30 reported lowest (36.6%). With regard to the familiarity of reflection and reflective practice, there was significant age effect. Group above 43 reported highest (56.3%), while Group 31–35 reported lowest (47.3%). As to the use of reflective learning methods, the age effect was significant. Group above 43 reported highest (56.4%), while Group 19–30 reported lowest (45.4%). For the familiarity of lifelong learning, there were more than 80% participants choosing '*Yes*' answer in all the four age groups. The significant age effect showed that Group above 43 reported highest (91.8%), while Group 19–30 reported lowest (85.3%) on '*Yes*' option. For the familiarity of co-operative learning, there were more than 60% participants choosing '*Yes*' answer in all the age groups. The age effect was significant. Group above 43 reported highest and Group 36–43 reported lowest. There were less than 50% participants choosing '*Yes*' answer in all the four age groups on the familiarity of portfolio building. The age effect was significant. Group above 43 reported highest and Group 19–31 reported lowest (see Table 5.12).

Influence of Profession

For the implementation of PBL in their universities, there were more than 70% participants choosing '*Yes*' answer in all the four profession groups. The differences among these profession groups was significant. Teaching administrators reported highest and Clinical teaching staff reported lowest. Concerning the

Table 5.12. Staff's knowledge about new teaching methods by age.

Item	19–30 N(%)		31–35 N(%)		36–43 N(%)		>43 N(%)		χ^2	*P*
Has PBL been implemented at your university?	303	(74.4)	326	(75.6)	326	(76.3)	313	(74.3)	0.63	0.889
Are you familiar with the concepts of mentoring and student support?	150	(36.6)	178	(41.3)	212	(49.4)	225	(53.1)	28.61	<0.001
Are you familiar with the concepts of reflection and reflective practice?	196	(47.8)	205	(47.3)	231	(53.2)	242	(56.3)	9.62	0.022
Do you use reflective learning methods in your teaching?	183	(45.4)	203	(47.7)	216	(50.7)	237	(56.4)	11.36	0.010
Are you familiar with the concept of lifelong learning?	348	(85.3)	379	(87.7)	380	(88.4)	393	(91.8)	8.87	0.031
Are you familiar with the concept of co-operative learning?	267	(64.8)	300	(69.6)	275	(63.7)	329	(76.5)	20.37	<0.001
Are you familiar with the concept of portfolio building?	138	(33.8)	151	(34.9)	180	(41.8)	204	(47.3)	21.36	<0.001

familiarity of mentoring and student support, there were less than 50% participants choosing '*Yes*' answer in all the four profession groups. Clinical teaching staff reported highest and Public health teaching staff reported lowest. With regard to the familiarity of reflection and reflective practice, there was significant profession effect. Clinical teaching staff reported highest (56.6%) and Public health teaching staff reported lowest (43.1%). As to the use of reflective learning methods, the profession effect was significant. Clinical teaching staff reported highest (56.2%), while Teaching administrators reported lowest (37.6%). For the familiarity of lifelong learning, there were more than 80% participants choosing '*Yes*' answer in all the four profession groups. There was no significant profession effect. For the familiarity of co-operative learning, there were more than 60% participants choosing '*Yes*' answer in all the profession groups. The profession effect was not significant. There were less than 50% participants choosing '*Yes*' answer in all the four profession groups on the familiarity of portfolio building. The difference among these job title groups was significant. Clinical teaching staff reported highest and Public health teaching staff reported lowest (see Table 5.13).

Table 5.13. Staff's knowledge about new teaching methods by profession.

Item	Teaching administrators N(%)		Basic studies teaching staff N(%)		Public health teaching staff N(%)		Clinical teaching staff N(%)		χ^2	*P*
Has PBL been implemented at your university?	110	(86.6)	555	(75.5)	172	(78.5)	514	(73.0)	11.96	0.008
Are you familiar with the concepts of mentoring and student support?	54	(42.5)	331	(44.9)	85	(39.4)	357	(49.9)	9.30	0.026
Are you familiar with the concepts of reflection and reflective practice?	68	(53.1)	368	(49.3)	94	(43.1)	406	(56.6)	15.27	0.002
Do you use reflective learning methods in your teaching?	44	(37.6)	375	(51.1)	91	(41.6)	397	(56.2)	23.73	< 0.001
Are you familiar with the concept of lifelong learning?	117	(92.1)	660	(89.1)	190	(87.2)	625	(87.8)	2.63	0.452
Are you familiar with the concept of co-operative learning?	88	(68.8)	503	(67.6)	155	(70.5)	506	(70.9)	1.98	0.576
Are you familiar with the concept of portfolio building?	51	(39.8)	272	(36.6)	59	(26.8)	345	(48.4)	40.22	< 0.001

Influence of Job Title

For the implementation of PBL in participants' universities, no matter whether their job title are junior or senior, more than 70% participants responded with '*Yes*'. The difference among these job title groups was not significant. Concerning the familiarity of mentoring and student support, there were significant difference among these three job title groups. Senior group reported highest (53.0%) and Junior group reported lowest (40.3%). With regard to the familiarity of reflection and reflective practice, there was significant job title effect. Senior group reported highest (56.3%) and Middle group reported lowest (47.4%). As to the use of reflective learning methods, the job title effect was not significant. For the familiarity of lifelong learning, there were more than 80% participants choosing '*Yes*' answer in all the three job title groups. There was no significant job title effect. For the familiarity of co-operative learning,

Table 5.14. Staff's knowledge about new teaching methods by job title.

Item	Junior N(%)		Middle N(%)		Senior N(%)		χ^2	P
Has PBL been implemented at your university?	323	(74.6)	457	(74.9)	571	(77.0)	1.12	0.570
Are you familiar with the concepts of mentoring and student support?	175	(40.3)	254	(41.6)	398	(53.0)	25.25	<0.001
Are you familiar with the concepts of reflection and reflective practice?	218	(49.9)	292	(47.4)	426	(56.3)	11.48	0.003
Do you use reflective learning methods in your teaching?	213	(49.7)	292	(48.3)	402	(54.0)	4.75	0.093
Are you familiar with the concept of lifelong learning?	380	(88.0)	534	(87.0)	678	90.2)	3.58	0.167
Are you familiar with the concept of co-operative learning?	307	(70.6)	405	(65.7)	540	(71.5)	5.74	0.057
Are you familiar with the concept of portfolio building?	180	(41.6)	204	(33.1)	343	(45.4)	21.76	<0.001

there were more than 60% participants choosing '*Yes*' answer in all the job title groups. The difference among the three job title groups was not significant. There were less than 50% participants choosing '*Yes*' answer in all the three job title groups on the familiarity of portfolio building. The significant among these job title groups was significant. Senior group reported highest and Middle group reported lowest (see Table 5.14).

Influence of University's Geographical Location

For the implementation of PBL in their universities, there were more than 60% participants choosing '*Yes*' answer in both East and West groups. Concerning the familiarity of mentoring and student support, both East and West groups reported less than 50%. With regard to the familiarity of reflection and reflective practice, East group reported 53.5%, while West group reported 44.4% on '*Yes*' option. As to the use of reflective learning methods, East group reported 52.4%, while West group reported 45.5%. For the familiarity of lifelong learning, there were more than 80% participants choosing '*Yes*' answer in both East and West group. For the familiarity of co-operative learning, there were more than 60% participants choosing '*Yes*' answer in both East and West group.

Table 5.15. Staff's knowledge about new teaching methods by location.

Item	East N/(%)		West N(%)		χ^2	*P*
Has PBL been implemented at your university?	1132	(78.8)	219	(62.8)	39.45	<0.001
Are you familiar with the concepts of mentoring and student support?	694	(48.1)	133	(37.7)	12.39	<0.001
Are you familiar with the concepts of reflection and reflective practice?	778	(53.5)	158	(44.4)	9.54	0.002
Do you use reflective learning methods in your teaching?	747	(52.4)	160	(45.5)	5.48	0.019
Are you familiar with the concept of lifelong learning?	1287	(89.3)	305	(85.4)	4.24	0.039
Are you familiar with the concept of co-operative learning?	1027	(70.9)	225	(63.0)	8.30	0.004
Are you familiar with the concept of portfolio building?	610	(42.1)	117	(33.0)	9.90	0.002

There were less than 50% participants choosing '*Yes*' answer in both East and West groups on the familiarity of portfolio building. The effect of university's geographical location was significant on all the seven questions, with East group reporting higher percentage than the West group (see Table 5.15).

5.3 Summary

On the purpose of investigating general opinions on staff's attitude and knowledge about new teaching methods in medicine and health education, there are two sections in this chapter. First, participants' ideas on teaching methods: to what extent they think alternative teaching methods (e.g. PBL) should be introduced in medicine and health education, what they think of the effect of PBL, to what extent they agree that reflective learning methods should be implemented in medical teaching, to what extent they think that the concept of lifelong learning should be integrated in the teaching and learning process. Second, staff's knowledge about new teaching methods in medicine and health education: Has PBL been implemented at their university? Are they familiar with the concepts of mentoring and student support? Are they familiar with the concepts of reflection and reflective practice? Do they use reflective learning methods in their teaching? Are they familiar with the concepts of lifelong learning, co-operative learning and portfolio building?

Eleven questions were asked with regard to these aspects in the survey that included four 5-point Likert-type questions and seven '*Yes/No*' questions.

Overall findings to these questions, and results analyzed according to gender, age, profession, job title, and university's geographical locations were also reported respectively in two sections in this chapter.

Overall findings in this chapter can be summarized in the following:

1. Participants in this investigation generally agreed on the idea and implementation of new teaching methods in medicine and health education. They thought that alternative teaching methods (e.g. PBL) should be introduced in medicine and health education. Staff who participated in this investigation held positive attitudes towards the effect of PBL. They also agree that reflective learning methods should be implemented in medical teaching. Participants believed that the concept of lifelong learning should be integrated in the teaching and learning process. These findings show a positive overall picture of staff's support for new teaching methods in medicine and health education. Significant differences can be observed between some groups, for example, Teaching administrators and Clinical teaching staff reported a higher degree of support for new methods than Basic teaching staff, and East group reported a higher degree of supporting educational reform than West group.
2. The majority of staff had knowledge about new teaching methods on most of the related topics. In general, there were more than 50% participants answering '*Yes*' on the following questions: the implementation of PBL in their universities, the familiarity of lifelong learning, and the familiarity of co-operative learning. There were two topics with which the majority of staff were not familiar. Generally speaking, less than 50% participants answered '*Yes*' on the familiarity of mentoring and student support, and the familiarity of portfolio building. These findings indicate that staff in Chinese medical universities have certain knowledge about the new teaching methods mentioned in this study, although significant differences were reported among the different methods. Considerate differences were also seen among different groups, for example, East group reported a higher degree of familiarity in all the methods than West group.

To conclude, a strong agreement was evidenced among the participants that new teaching methods should be employed in the current medicine and health education in China. Staff who have participated in the present investigation reported they possess certain level of knowledge about these new teaching methods, with differences of the familiarity level in some aspects due to factors such as age, profession, job title and geographical location of their university. Nevertheless, considering the fact that the pilot stage of reforms have already been conducted in these universities, the respondent staff reported a limited level of knowledge about the new teaching methods. This indicates further need for facilitation with the staff's knowledge about and skills in implementing these new methods.

6

Assessment Methods Reform — Teaching Staff's Attitudes and Participation

Xiaoju Duan, Jiannong Shi and Xiangyun Du

Abstract

This chapter aims to investigate staff's knowledge about and attitudes towards assessment in medicine and health education in China in general, as well as certain new forms of assessments in particular. Sixteen questions were asked with regard to these aspects in the survey. Findings are presented and analyzed in this chapter, which are structured in five sections. Section 1 reports participants' responses to (1) to what extent they agree that teaching evaluations are an important part of the teaching and learning process; (2) to what extent they think that there should be an assessment of basic medical knowledge for students who have taken the relevant courses after their second and third years of study; (3) to what extent they believe that there should be an assessment of medical practice before students graduate; (4) to what extent they believe that the Standard Patient (SP) exam should be used in the medical practice assessment for graduation; (5) to what extent they think that exam results should be centrally analyzed and sent back to relevant departments afterwards. Section 2 presents participants' responses to methods which should be used to evaluate teaching. Section 3 reports the evaluation methods which have been used at

Change and Reform in Medicine and Health Education in China, 87–116.

their universities. Section 4 provides an overview picture of the computer-based medical test database in participants' universities. Section 5 shows the overall situations about medical universities' exams and staff's familiarity about it. Data analyses in this chapter not only include overall findings to the above addressed issues, but also take into consideration the impact of gender, age, profession, job title, and the university's geographical location, respectively.

6.1 General Opinions on Educational Assessment

In order to understand staff's overall perception on the assessment in medicine and health education, five questions were asked that require 5-point Likert-type answers. These questions include: (1) To what extent did they agree that teaching evaluations are an important part of the teaching and learning process (see survey Q27 Appendix 1)? (2) To what extent did they think there should be an assessment of basic medical knowledge for students who have taken the relevant courses after their second and third years of study (see survey Q31 Appendix 1)? (3) To what extent did they believe there should be an assessment of medical practice before students graduate (see survey Q33 Appendix 1)? (4) To what extent did they believe that the Standard Patient (SP) exam should be used in the medical practice assessment for graduation (see survey Q36 Appendix 1)? And (5) To what extent did they think that exam results should be analyzed centrally and sent back to relevant departments afterwards (see survey Q40 Appendix 1)?

Overall findings

The overall descriptive analysis results of participants' opinions on the importance of teaching evaluations in the teaching and learning process, assessment of basic medical knowledge, assessment of medical practice before students graduate, use of Standard Patient (SP) exam, analysis and feedback of exam results, are presented in Table 6.1. In general, participants reported higher than the mean score on all these five questions.

Influence of Gender

Both male and female groups held positive attitude towards educational assessment in medicine and health. They had similar opinions on the importance of

Table 6.1. General opinions on assessment.

Item	*N*	*M*	*SD*
To what extent do you agree that teaching evaluations are an important part of the teaching and learning process?	1692	4.29	0.72
To what extent do you think there should be an assessment of basic medical knowledge for students who have taken the relevant courses after their 2nd and 3rd years of study?	1783	4.12	0.86
To what extent do you believe there should be an assessment of medical practice before students graduate?	1783	4.31	0.76
To what extent do you believe that the Standard Patient (SP) exam should be used on the medical practice assessment for graduation?	1747	4.17	0.80
To what extent do you think that exam results should be analyzed centrally and sent back to relevant departments afterwards?	1768	4.33	0.67

teaching evaluations in the teaching and learning process, assessment of basic medical knowledge, assessment of medical practice before students graduate, and analysis and feedback of exam results. Males reported higher than females on the use of Standard Patient (SP) exam (see Table 6.2).

Table 6.2. General opinions on assessment by gender.

Item	Gender	*N*	*M*	*SD*	*t*	*P*
To what extent do you agree that teaching evaluations are an important part of the teaching and learning process?	Male	749	4.32	0.73	1.56	0.120
	Female	907	4.27	0.70		
To what extent do you think there should be an assessment of basic medical knowledge for students who have taken the relevant courses after their 2nd and 3rd years of study?	Male	791	4.10	0.89	−0.72	0.474
	Female	951	4.13	0.83		
To what extent do you believe there should be an assessment of medical practice before students graduate?	Male	791	4.33	0.76	0.72	0.475
	Female	952	4.30	0.76		
To what extent do you believe that the Standard Patient (SP) exam should be used on the medical practice assessment for graduation?	Male	777	4.22	0.79	2.44	0.015
	Female	932	4.13	0.80		
To what extent do you think that exam results should be analyzed centrally and sent back to relevant departments afterwards?	Male	784	4.34	0.66	0.20	0.841
	Female	944	4.33	0.67		

Influence of Age

In general, all the four age groups were positive to the new assessment in medicine and health education. Significant age effect was found with regard to the importance of teaching evaluations in the teaching and learning process. Participants of different age groups had similar opinions on the assessment of basic medical knowledge, assessment of medical practice before students graduate, use of Standard Patient (SP) exam, and analysis and feedback of exam results (see Table 6.3).

Table 6.3. General opinions on assessment by age.

Item	Age group	*N*	*M*	*SD*	*F*	*P*
To what extent do you agree that teaching evaluations are an important part of the teaching and learning process?	19–30	383	4.22	0.72	3.18	0.023
	31–35	412	4.27	0.66		
	36–43	392	4.32	0.75		
	>43	409	4.37	0.67		
To what extent do you think there should be an assessment of basic medical knowledge for students who have taken the relevant courses after their 2nd and 3rd years of study?	19–30	409	4.13	0.82	0.56	0.641
	31–35	426	4.10	0.87		
	36–43	422	4.14	0.83		
	> 43	425	4.07	0.90		
To what extent do you believe there should be an assessment of medical practice before students graduate?	19–30	409	4.30	0.82	0.02	0.997
	31–35	425	4.32	0.71		
	36–43	425	4.31	0.78		
	>43	423	4.31	0.75		
To what extent do you believe that the Standard Patient (SP) exam should be used on the medical practice assessment for graduation?	19–30	403	4.14	0.83	1.16	0.323
	31–35	420	4.14	0.81		
	36–43	416	4.18	0.77		
	>43	411	4.23	0.78		
To what extent do you think that exam results should be analyzed centrally and sent back to relevant departments afterwards?	19–30	406	4.32	0.66	1.05	0.368
	31–35	421	4.33	0.65		
	36–43	422	4.31	0.68		
	>43	419	4.38	0.69		

Post hoc tests showed the *P*-values of comparisons between any two age groups (see Table 6.4). Group 19–30 reported lower than Group 36–43 and Group above 43 on the importance of teaching evaluations in the teaching and learning process. Group 31–35 evaluated lower than Group above 43 on the importance of teaching evaluations in the teaching and learning process.

Influence of Profession

Impact of profession was significant on all aspects of assessments in medicine and health education. Participants of different professions had different opinions on the importance of teaching evaluations in the teaching and learning process, assessment of basic medical knowledge, assessment of medical practice before students graduate, use of Standard Patient (SP) exam, and in the analysis and feedback of exam results (see Table 6.5).

Table 6.4. General opinions on assessment by age-post hoc analysis.

Item	19–30 vs 31–35	19–30 vs 36–43	19–30 vs >43	31–35 vs 36–43	31–35 vs >43	36–43 vs >43
To what extent do you agree that teaching evaluations are an important part of the teaching and learning process?	0.342	0.048	0.004	0.292	0.047	0.361
To what extent do you think there should be an assessment of basic medical knowledge for students who have taken the relevant courses after their 2nd and 3rd years of study?	0.601	0.959	0.283	0.562	0.578	0.257
To what extent do you believe there should be an assessment of medical practice before students graduate?	0.819	0.924	0.902	0.893	0.915	0.978
To what extent do you believe that the Standard Patient (SP) exam should be used on the medical practice assessment for graduation?	0.952	0.460	0.120	0.420	0.102	0.408
To what extent do you think that exam results should be analyzed centrally and sent back to relevant departments afterwards?	0.831	0.795	0.170	0.632	0.243	0.100

The P-values of post hoc tests between any two profession groups are shown in Table 6.6. Teaching administrators reported higher than the other three professions on the importance of teaching evaluations in the teaching and learning process. Teaching administrators, Basic studies teaching staff, and Clinical teaching staff reported higher than Public health teaching staff on the assessment of basic medical knowledge. Clinical teaching staff evaluated higher than Basic studies teaching staff on the assessment of basic medical knowledge. Teaching administrators and Clinical teaching staff reported higher than basic studies teaching staff, which reported higher than Public health teaching staff on the assessment of medical practice before students graduate and the use of Standard Patient (SP) exam. Teaching administrators reported higher than Basic studies teaching staff, Public health teaching staff, and Clinical teaching staff on the analysis and feedback of exam results.

Table 6.5. General opinions on assessment by profession.

Item	Profession	*N*	*M*	*SD*	*F*	*P*
To what extent do you agree that teaching evaluations are an important part of the teaching and learning process?	Teaching administrators	124	4.52	0.74	5.58	0.001
	Basic studies teaching staff	690	4.24	0.75		
	Public health teaching staff	210	4.29	0.69		
	Clinical teaching staff	668	4.29	0.68		
To what extent do you think there should be an assessment of basic medical knowledge for students who have taken the relevant courses after their 2nd and 3rd years of study?	Teaching administrators	128	4.24	0.87	18.25	<0.001
	Basic studies teaching staff	731	4.03	0.88		
	Public health teaching staff	220	3.86	0.90		
	Clinical teaching staff	704	4.27	0.78		
To what extent do you believe there should be an assessment of medical practice before students graduate?	Teaching administrators	128	4.48	0.78	13.64	<0.001
	Basic studies teaching staff	727	4.25	0.75		
	Public health teaching staff	221	4.10	0.84		
	Clinical teaching staff	707	4.41	0.72		
To what extent do you believe that the Standard Patient (SP) exam should be used on the medical practice assessment for graduation?	Teaching administrators	128	4.41	0.82	29.06	<0.001
	Basic studies teaching staff	709	4.04	0.80		
	Public health teaching staff	216	3.92	0.83		
	Clinical teaching staff	694	4.34	0.73		
To what extent do you think that exam results should be analyzed centrally and sent back to relevant departments afterwards?	Teaching administrators	128	4.58	0.67	7.50	<0.001
	Basic studies teaching staff	719	4.28	0.68		
	Public health teaching staff	219	4.30	0.63		
	Clinical teaching staff	702	4.34	0.67		

Influence of Job Title

All the three job title groups were positive about these five questions, however, the impact of job title was significant on the importance of teaching evaluations in the teaching and learning process and marginally significant on the assessment of basic medical knowledge. There was no impact of job titles on the assessment of medical practice before students graduate, use of Standard Patient (SP) exam, and the analysis and feedback of exam results (see Table 6.7).

The P-values of post hoc tests between any two title groups are shown in Table 6.8. Senior group reported higher than Junior and Middle groups on the importance of teaching evaluations in the teaching and learning process.

Influence of University's Geographical Location

Both East and West groups held positive attitudes towards new assessment in medicine and health education. They had similar opinions on the importance

Table 6.6. General opinions on assessment by profession — post hoc analysis.

Item	P1 vs P2	P1 vs P3	P1 vs P4	P2 vs P3	P2 vs P4	P3 vs P4
To what extent do you agree that teaching evaluations are an important part of the teaching and learning process?	<0.001	0.003	0.001	0.421	0.172	0.891
To what extent do you think there should be an assessment of basic medical knowledge for students who have taken the relevant courses after their 2nd and 3rd years of study?	0.009	<0.001	0.719	0.008	<0.001	<0.001
To what extent do you believe there should be an assessment of medical practice before students graduate?	0.001	<0.001	0.285	0.006	<0.001	<0.001
To what extent do you believe that the Standard Patient (SP) exam should be used on the medical practice assessment for graduation?	<0.001	<0.001	0.356	0.045	<0.001	<0.001
To what extent do you think that exam results should be analyzed centrally and sent back to relevant departments afterwards?	<0.001	<0.001	<0.001	0.635	0.081	0.469

Note: P1, Teaching administrators; P2, Basic studies teaching staff; P3, Public health teaching staff, P4,Clinical teaching staff.

Table 6.7. General opinions on assessment by job title.

Item	Job title	*N*	*M*	*SD*	*F*	*P*
To what extent do you agree that teaching evaluations are an important part of the teaching and learning process?	Junior	404	4.24	0.72	8.66	<0.001
	Middle	571	4.22	0.75		
	Senior	717	4.37	0.68		
To what extent do you think there should be an assessment of basic medical knowledge for students who have taken the relevant courses after their 2nd and 3rd years of study?	Junior	431	4.19	0.83	2.88	0.056
	Middle	601	4.06	0.85		
	Senior	751	4.13	0.87		
To what extent do you believe there should be an assessment of medical practice before students graduate?	Junior	431	4.34	0.75	1.39	0.250
	Middle	603	4.27	0.80		
	Senior	749	4.33	0.73		
To what extent do you believe that the Standard Patient (SP) exam should be used on the medical practice assessment for graduation?	Junior	426	4.15	0.82	2.64	0.072
	Middle	591	4.12	0.80		
	Senior	730	4.22	0.77		
To what extent do you think that exam results should be analyzed centrally and sent back to relevant departments afterwards?	Junior	429	4.28	0.70	2.54	0.079
	Middle	597	4.30	0.65		
	Senior	742	4.37	0.67		

Table 6.8. General opinions on assessment by job title — post hoc analysis.

Item	Junior vs Middle	Junior vs Senior	Middle vs Senior
To what extent do you agree that teaching evaluations are an important part of the teaching and learning process?	0.557	0.004	<0.001
To what extent do you think there should be an assessment of basic medical knowledge for students who have taken the relevant courses after their 2nd and 3rd years of study?	0.018	0.245	0.146
To what extent do you believe there should be an assessment of medical practice before students graduate?	0.153	0.845	0.152
To what extent do you believe that the Standard Patient (SP) exam should be used on the medical practice assessment for graduation?	0.606	0.142	0.027
To what extent do you think that exam results should be analyzed centrally and sent back to relevant departments afterwards?	0.631	0.041	0.089

of teaching evaluations in the teaching and learning process, and the use of Standard Patient (SP) exam. East group reported higher than West group on the assessment of basic medical knowledge, assessment of medical practice before students graduate, and the analysis and feedback of exam results (see Table 6.9).

Table 6.9. General opinions on assessment by university's geographical location.

Item	Location	*N*	*M*	*SD*	*t*	*P*
To what extent do you agree that teaching evaluations are an important part of the teaching and learning process?	East	1372	4.29	0.71	0.70	0.482
	West	320	4.26	0.72		
To what extent do you think there should be an assessment of basic medical knowledge for students who have taken the relevant courses after their 2nd and 3rd years of study?	East	1438	4.15	0.85	3.39	0.001
	West	345	3.98	0.86		
To what extent do you believe there should be an assessment of medical practice before students graduate?	East	1436	4.34	0.75	3.18	0.001
	West	347	4.20	0.78		
To what extent do you believe that the Standard Patient (SP) exam should be used on the medical practice assessment for graduation?	East	1409	4.18	0.80	1.53	0.127
	West	338	4.11	0.77		
To what extent do you think that exam results should be analyzed centrally and sent back to relevant departments afterwards?	East	1425	4.35	0.67	2.49	0.013
	West	343	4.24	0.70		

6.2 Methods to Evaluate Teaching

In order to understand staff's overall perception on methods being used to evaluate teaching, one multiple choice question was asked with four choices (see survey Q28 Appendix 1). The choices were: (1) Evaluation by management staff; (2) Evaluation by peers; (3) Evaluation by self; and (4) Evaluation by students.

Overall findings

Participants' suggestions on the methods which should be used to evaluate teaching are shown in Table 6.10 ($N = 1788$). Less than 50% participants selected the choices of Evaluation by management staff. More than 50% participants selected the choices of Evaluation by peers, Evaluation by one's self and Evaluation by students. Results of participants' responses on the four choices can be ranked as following: Evaluation by peers and Evaluation by students, Evaluation by one's self and Evaluation by management staff.

Influence of Gender

Both male and female groups reported around 50% on Evaluation by management staff, and the difference between male and female groups was not significant. More than 80% participants chose Evaluation by peers in both male and female groups. The difference between male and female group was not significant. Male group (47.3%) reported less percentage than female group (52.7%) on the choice of Evaluation by one's self. Both male and female groups reported more than 80% on the choice of Evaluation by students and there was no significant difference between male and female groups (see Table 6.11).

The male group ranked the choices in the order of Evaluation by peers, Evaluation by students, Evaluation by management staff, and Evaluation by

Table 6.10. Which methods should be used to evaluate teaching.

Item	Yes N(%)	No N(%)
Evaluation by management staff	859 (48.0)	929 (52.0)
Evaluation by peers	1466 (82.0)	322 (18.0)
Evaluation by one's self	899 (50.3)	889 (49.7)
Evaluation by students	1466 (82.0)	322 (18.0)

Table 6.11. Which methods should be used to evaluate teaching by gender.

Item	Male N(%)		Female N(%)		χ^2	*P*
Evaluation by management staff	401	(50.6)	442	(46.3)	3.28	0.070
Evaluation by peers	644	(81.3)	787	(82.4)	0.35	0.554
Evaluation by one'sself	375	(47.3)	503	(52.7)	4.91	0.027
Evaluation by students	644	(81.3)	785	(82.2)	0.23	0.633

one's self. The female group ranked the choices in the order of Evaluation by peers, Evaluation by students, Evaluation by one's self, and Evaluation by management staff.

Influence of Age

For the choice of Evaluation by management staff, Group 19–30 and Group 31–35 reported less than 50%, and Group 36–43 and Group above 43 reported more than 50%. The differences among these four age groups was significant. For the choice of Evaluation by peers, all the age groups reported more than 70%. The age effect was significant on this choice with Group above 43 reporting highest and Group 19–30 reported the lowest. For the choice of Evaluation by one's self, all the four age groups reported around 50%, and the difference between groups was not significant. For the choice of Evaluation by students, there were more than 80% participants choosing it in all these four age groups. The difference among age groups was not significant.

The choice rankings of the four age groups were slightly different, but the general trend was that Evaluation by peers and Evaluation by students was always the highest, and Evaluation by management staff and Evaluation by one's self was always the lowest (see Table 6.12).

Table 6.12. Which methods should be used to evaluate teaching by age.

Item	19–30 N(%)		31–35 N(%)		36–43 N(%)		>43 N(%)		χ^2	*P*
Evaluation by management staff	174	(42.9)	183	(43.0)	226	(52.9)	233	(54.4)	19.80	<0.001
Evaluation by peers	313	(77.1)	355	(83.3)	351	(82.2)	366	(85.5)	10.76	0.013
Evaluation by one's self	200	(49.3)	210	(49.3)	212	(49.6)	223	(52.1)	0.95	0.814
Evaluation by students	320	(78.8)	344	(80.8)	355	(83.1)	365	(85.3)	6.74	0.081

Influence of Profession

For the choice of Evaluation by management staff, Teaching administrators and Clinical teaching staff reported more than 50%, and Basic studies teaching staff and Public health teaching staff reported less than 50%. The differences among these four profession groups were significant. For the choice of Evaluation by peers, all the profession groups reported more than 70%. Impact of profession was not significant on this choice. For the choice of Evaluation by one's self, there was a significant profession effect. Teaching administrators reported highest (66.4%) and Basic studies teaching staff reported lowest (46.3%). For the choice of Evaluation by students, there were more than 70% participants choosing it in all these four profession groups. The difference among profession groups was significant, with Teaching administrators reporting highest and Basic studies teaching staff reporting lowest.

The choice rankings of the four profession groups were slightly different. Teaching administrators and Clinical teaching staff had the same choice ranking: Evaluation by students, Evaluation by peers, Evaluation by management staff and Evaluation by one's self. Basic studies teaching staff and Public health teaching staff had the same choice ranking: Evaluation by peers, Evaluation by students, Evaluation by one's self , and Evaluation by management staff (see Table 6.13).

Influence of Job Title

For the choice of Evaluation by management staff, there was significant difference among the three job title groups, with Senior group reporting highest

Table 6.13. Which methods should be used to evaluate teaching by profession.

Item	Teaching administrators N(%)		Basic studies teaching staff N(%)		Public health teaching staff N(%)		Clinical teaching staff N(%)		χ^2	P
Evaluation by management staff	93	(72.7)	286	(39.1)	100	(45.2)	380	(53.7)	64.58	<0.001
Evaluation by peers	111	(86.7)	603	(82.4)	190	(86.0)	562	(79.5)	7.38	0.061
Evaluation by one's self	85	(66.4)	339	(46.3)	115	(52.0)	360	(50.9)	18.32	<0.001
Evaluation by students	116	(90.6)	577	(78.8)	186	(84.2)	587	(83.0)	12.65	0.005

(52.6%) and Middle group reported lowest (44.1%). For the choice of Evaluation by peers, all the job title groups reported more than 70%. The job title effect was significant on this choice, with Senior group reporting highest and Junior group reporting lowest. For the choice of Evaluation by one's self, all these three job title groups reported around 50% and there was no significant job title effect. For the choice of Evaluation by students, there were more than 70% participants choosing it in all these three job title groups. The difference among job title groups was significant, with Senior group reporting highest and Middle group reporting lowest.

The choice rankings of the three job title groups were slightly different, but there was a general trend with Evaluation by peers and Evaluation by students always scoring the highest, and Evaluation by management staff and Evaluation by one's self were always the lowest (see Table 6.14).

Table 6.14. Which methods should be used to evaluate teaching by job title.

Item	Junior N(%)		Middle N(%)		Senior N(%)		χ^2	*P*
Evaluation by management staff	196	(45.7)	267	(44.1)	396	(52.6)	11.04	0.004
Evaluation by peers	329	(76.7)	491	(81.0)	646	(85.8)	15.91	<0.001
Evaluation by one's self	215	(50.1)	297	(49.0)	387	(51.4)	0.77	0.681
Evaluation by students	344	(80.2)	482	(79.5)	640	(85.0)	8.01	0.018

Influence of University's Geographical Location

East group (46.8%) reported less percentage than West group (53.0%) did on the choice of Evaluation by management staff. There were more than 80% participants choosing Evaluation by peers in both East and West groups. The difference between East and West group was not significant. East and West groups reported around 50% on the choice of Evaluation by one's self and there was no significant difference between East and West groups. Both East and West groups reported more than 80% on the choice of Evaluation by students and there was no significant difference between East and West groups (see Table 6.15).

The choices ranking of the East group was Evaluation by students, Evaluation by peers, Evaluation by one's self, and Evaluation by management staff. The choices ranking of the West group was Evaluation by peers, Evaluation by students, Evaluation by one's self, and Evaluation by management staff.

Table 6.15. Which methods should be used to evaluate teaching by university geographical location.

Item	East N(%)		West N(%)		χ^2	P
Evaluation by management staff	674	(46.8)	185	(53.0)	4.28	0.038
Evaluation by peers	1178	(81.9)	288	(82.5)	0.08	0.774
Evaluation by one's self	712	(49.5)	187	(53.6)	1.89	0.169
Evaluation by students	1184	(82.3)	282	(80.8)	0.42	0.519

6.3 What Kinds of Evaluation Methods Have Been Used

In order to understand staff's overall perception on the kinds of evaluation methods which have been used at their universities, one multiple choice question was asked with four choices to select (see survey Q29 Appendix 1). The choices were: (1) Evaluation by management staff; (2) Evaluation by peers; (3) Evaluation by selves; and (4) Evaluation by students.

Overall findings

Participants' responses on what kinds of evaluation methods have been used at their universities are shown in Table 6.16 ($N = 1775$). There were more than 50% participants selecting the choices of Evaluation by management staff, Evaluation by peers, and Evaluation by students. There were less than 50% participants selecting the choices of Evaluation by selves. Results of participants' selections on the four choices can be ranked as following: Evaluation by students, Evaluation by peers, Evaluation by management staff, and Evaluation by selves.

Influence of Gender

Both male and female groups reported more than 50% on Evaluation by management staff, with male group reporting significantly higher than female

Table 6.16. What kinds of evaluation methods have been used.

Item	Yes N(%)	No N(%)
Evaluation by management staff	1156 (65.1)	619 (34.9)
Evaluation by peers	1252 (70.5)	523 (29.5)
Evaluation by selves	725 (40.8)	1050 (59.2)
Evaluation by students	1509 (85.0)	266 (15.0)

group. More than 60% participants chose Evaluation by peers in both male and female group. The difference between male and female group was not significant. Both male and female groups reported less than 50% on the choice of Evaluation by selves, and the difference between male and female groups was not significant. Both male and female groups reported more than 80% on the choice of Evaluation by students with males reporting less than females (see Table 6.17).

Table 6.17. What kinds of evaluation methods have been used by gender.

Item	Male N(%)		Female N(%)		χ^2	P
Evaluation by management staff	534	(67.8)	597	(63.0)	4.35	0.037
Evaluation by peers	567	(72.0)	655	(69.1)	1.69	0.194
Evaluation by selves	335	(42.5)	373	(39.3)	1.79	0.181
Evaluation by students	649	(82.4)	824	(86.9)	6.96	0.008

The choices rankings of the male and female groups were the same: Evaluation by students, Evaluation by peers, Evaluation by management staff, and Evaluation by selves.

Influence of Age

For the choice of Evaluation by management staff, all the age groups reported more than 60%. The differences among these four age groups were not significant. For the choice of Evaluation by peers, all the age groups reported more than 60%. The age differences were significant on this choice with Group above 43 reported highest and Group 19–30 reported the lowest. For the choice of Evaluation by selves, all the four age groups reported less than 50%. The differences between groups were significant, with Group above 43 reported highest and Group 31-35 reported lowest. For the choice of Evaluation by students, there were more than 80% participants choosing it in all these four age groups. The difference among age groups was significant, with Group above 43 reported highest and Group 19-30 reported lowest.

The choice ranking of Group 19–30 was slightly different from other three age groups. The ranking of Group 19–30 was Evaluation by students, Evaluation by management staff, Evaluation by peers, and Evaluation by selves. The ranking of the other three age groups was Evaluation by management staff, Evaluation by students, Evaluation by peers, and Evaluation by selves (see Table 6.18).

Table 6.18. What kinds of evaluation methods have been used by age.

Item	19–30 N(%)		31–35 N(%)		36–43 N(%)		>43 N(%)		χ^2	P
Evaluation by management staff	263	(65.3)	272	(64.0)	282	(66.4)	280	(66.4)	0.71	0.872
Evaluation by peers	256	(63.5)	291	(68.5)	309	(72.7)	325	(77.0)	19.88	<0.001
Evaluation by selves	158	(39.2)	155	(36.5)	179	(42.1)	198	(46.9)	10.43	0.015
Evaluation by students	324	(80.4)	367	(86.4)	363	(85.4)	371	(87.9)	10.21	0.017

Influence of Profession

For the choice of Evaluation by management staff, all the four profession groups reported more than 50%. The difference among profession groups was significant, with Teaching administrators reporting highest, and Public health teaching staff reporting lowest. More than 60% of the all the profession groups chose Evaluation by peers. The profession effect was not significant on this choice. For the choice of Evaluation by selves, all profession groups reported less than 50%, and there was no significant profession effect. For the choice of Evaluation by students, more than 80% participants chose it in all these four profession groups. The difference among profession groups was significant, with Public health teaching staff reporting highest and Clinical teaching staff reporting lowest.

The choice rankings of Teaching administrators was slightly different from the other three profession groups. The ranking for Teaching administrators was Evaluation by students, Evaluation by management staff, Evaluation by peers, and Evaluation by selves. The ranking for the other three profession groups was Evaluation by students, Evaluation by peers, Evaluation by management staff, and Evaluation by selves (see Table 6.19).

Table 6.19. What kinds of evaluation methods have been used by profession.

Item	Teaching administrators N(%)		Basic studies teaching staff N(%)		Public health teaching staff N(%)		Clinical teaching staff N(%)		χ^2	P
Evaluation by management staff	103	(83.1)	444	(60.9)	132	(59.7)	477	(68.0)	28.75	<0.001
Evaluation by peers	92	(74.2)	515	(70.6)	148	(67.0)	497	(70.9)	2.20	0.532
Evaluation by selves	57	(46.0)	289	(39.6)	80	(36.2)	299	(42.7)	4.71	0.195
Evaluation by students	112	(90.3)	622	(85.3)	203	(91.9)	572	(81.6)	17.34	0.001

Influence of Job Title

For the choice of Evaluation by management staff, all the three job title groups reported more than 60%, and there was no significant differences among these groups. For the choice of Evaluation by peers, all the job title groups reported more than 60%. The job title effect was significant on this choice, with Senior group reporting highest and Junior group reporting lowest. For the choice of Evaluation by selves, all these three job title groups reported less than 50%. The significant job title effect indicated that Senior group reported highest and Middle group reported lowest. For the choice of Evaluation by students, more than 70% participants chose it in all these three job title groups. The difference among job title groups was significant, with Senior group reporting highest and Junior group reporting lowest.

The choice rankings of Junior group was different from the other two job title groups. Junior group's ranking was in the order of: Evaluation by students, Evaluation by management staff, Evaluation by peers, and Evaluation by selves. The ranking for the other two job title groups was Evaluation by students, Evaluation by peers, Evaluation by management staff, and Evaluation by selves (see Table 6.20).

Influence of University's Geographical Location

Both East and West groups reported more than 60% on the choice of Evaluation by management staff, and there was no significant difference between these two groups. More than 60% participants chose Evaluation by peers in both East and West groups. The difference between East and West groups was significant, and West group reported more than East group. For the choice of Evaluation by selves, East group (37.7%) reported lower than West group (54.1%). Both East and West groups reported more than 80% on the choice of Evaluation by students and there was no significant difference between East and West groups (see Table 6.21).

Table 6.20. What kinds of evaluation methods have been used by job title.

Item	Junior N(%)		Middle N(%)		Senior N(%)		χ^2	*P*
Evaluation by management staff	285	(66.6)	374	(62.2)	497	(66.6)	3.36	0.187
Evaluation by peers	277	(64.7)	420	(69.9)	555	(74.4)	12.44	0.002
Evaluation by selves	184	(43.0)	208	(34.6)	333	(44.6)	14.93	0.001
Evaluation by students	337	(78.7)	511	(85.0)	661	(88.6)	20.79	<0.001

Table 6.21. What kinds of evaluation methods have been used by university's geographical location.

Item	East N(%)		West N(%)		χ^2	*P*
Evaluation by management staff	936	(65.4)	220	(64.0)	0.26	0.611
Evaluation by peers	961	(67.2)	291	(84.6)	40.57	<0.001
Evaluation by selves	539	(37.7)	186	(54.1)	30.89	<0.001
Evaluation by students	1223	(85.5)	286	(83.1)	1.18	0.278

East group ranked their choices in the order of: Evaluation by students, Evaluation by peers, Evaluation by management staff, and Evaluation by selves. The items ranking of the West group was Evaluation by peers, Evaluation by students, Evaluation by management staff, and Evaluation by selves.

6.4 The Use of Computer-Based Medical Test Database

In order to understand the overall picture of the computer-based medical test database in participants' universities, a single-choice question was asked that required an option from *'Do not know'*, *'No'*, and *'Yes'* (see survey Q30 Appendix 1).

Overall findings

The descriptive data of participants' options on the use of computer-based medical test database for student assessment in their universities is shown in Table 6.22 ($N = 1785$). In general, most participants reported that their universities use computer-based medical test databases for student assessment.

Influence of Gender

Most male and female participants reported that their universities used computer-based medical test database for student assessment. Males and females reported similar percentages on the options of the use of computer-based medical test database for student assessment in their universities (see Table 6.23).

Table 6.22. The use of computer-based medical test database.

Item	Do not know N(%)	No N(%)	Yes N(%)
Does your university use a computer-based medical test database for student assessment?	249 (13.9)	268 (15.0)	1268 (71.0)

Table 6.23. The use of computer-based medical test database by gender.

Item	Male N(%)		Female N(%)		χ^2	*P*
Do not know	95	(12.0)	148	(15.5)	4.89	0.087
No	116	(14.7)	147	(15.4)		
Yes	578	(73.3)	660	(69.1)		

Influence of Age

Most participants chose *'Yes'* on this question in all the four age groups. Participants of these four age groups reported similarly on options of the use of computer-based medical test database for student assessment in their universities (see Table 6.24).

Table 6.24. The use of computer-based medical test database by age.

Item	19–30 N(%)		31–35 N(%)		36–43 N(%)		>43 N(%)		χ^2	*P*
Do not know	69	(16.9)	53	(12.5)	57	(13.4)	60	(14.1)	4.590	0.597
No	55	(13.5)	69	(16.3)	68	(16.0)	63	(14.8)		
Yes	284	(69.6)	301	(71.2)	301	(70.7)	303	(71.1)		

Influence of Profession

More than 50% participants chose *'Yes'* on this question in all these four profession groups. Participants of different professions reported different percentages on the options of the use of computer-based medical test database for student assessment in their universities (see details in Table 6.25). Public health teaching staff reported highest percentage on the options of *'Do not know'* and *'No'*. Clinical teaching staff reported highest on the option of *'Yes'* indicating their university did use the computer-based medical test database for student assessment.

Table 6.25. The use of computer-based medical test database by profession.

Item	Teaching administrators N(%)		Basic studies teaching staff N(%)		Public health teaching staff N(%)		Clinical teaching staff N(%)		χ^2	*P*
Do not know	10	(7.8)	128	(17.5)	60	(27.3)	51	(7.2)	119.54	<0.001
No	19	(14.8)	136	(18.6)	46	(20.9)	67	(9.5)		
Yes	99	(77.3)	466	(63.8)	114	(51.8)	589	(83.3)		

Influence of Job Title

Most participants reported highest on the option of '*Yes*' in all these three job title groups. There was no significant difference among these three job title groups on their options of the use of computer-based medical test database for student assessment in their universities (see details in Table 6.26).

Table 6.26. The use of computer-based medical test database by job title.

Item	Junior N(%)		Middle N(%)		Senior N(%)		χ^2	P
Do not know	69	(16.0)	81	(13.5)	99	(13.2)	4.16	0.385
No	58	(13.5)	101	(16.8)	109	(14.5)		
Yes	304	(70.5)	420	(69.8)	544	(72.3)		

Influence of University's Geographical Location

More than 50% participants chose '*Yes*' in both East and West groups. There was significant difference between East and West groups on the options. East group reported higher than West group on the '*Yes*' option, while West group reported higher on the '*Do not know*' and '*No*' options (see Table 6.27).

Table 6.27. The use of computer-based medical test database by university's geographical location.

Item	East N(%)		West N(%)		χ^2	P
Do not know	193	(13.4)	56	(16.1)	41.72	<0.001
No	180	(12.5)	88	(25.4)		
Yes	1065	(74.1)	203	(58.5)		

6.5 Universities' Exams and Staff's Familiarity About Them

This section gives an overview about medical universities' exams and staff's familiarity about it. Eight questions were asked that required '*Yes*' or '*No*' answers. These questions included: (1) Do their universities organize examinations on basic medical knowledge (see survey Q32 Appendix 1)? (2) Do their universities organize an examination on medical practice before students graduate (see survey Q34 Appendix 1)? (3) Are they familiar with the Standard Patient (SP) exam (see survey Q35 Appendix 1)? (4) Do their universities organize Standard Patient (SP) exams (see survey Q37 Appendix 1)? (5) Are they familiar with computer-based case simulation (CCS) examination (see survey Q38 Appendix 1)? (6) Do their universities use computer-based case simulation (CCS) examinations (see survey Q39 Appendix 1)? (7) Do their universities analyze exam results centrally and send them back to relevant departments afterwards (see survey Q41 Appendix 1)? (8) Are they familiar with portfolio assessments (see survey Q42 Appendix 1)?

Overall findings

The descriptive analysis results of participants' options on the exam on basic medical knowledge, the exam on medical practice before students graduate, the familiarity of Standard Patient (SP) exam, the organization of Standard Patient (SP) exam, the familiarity of computer-based case simulation (CCS) exam, the use of computer-based case simulation (CCS) examinations, the analysis and feedback of exam results, and the familiarity of portfolio assessment, are shown in Table 6.28. In general, more than 50% participants answered '*Yes*' on the exam on basic medical knowledge, the exam on medical practice before students graduate, the familiarity of Standard Patient (SP) exam, the familiarity of computer-based case simulation (CCS) exam, and the analysis and feedback of exam results. Less than 50% participants answered '*Yes*' on the organization of Standard Patient (SP) exam, the use of computer-based case simulation (CCS) examinations, and the familiarity of portfolio assessment.

Table 6.28. Universities' exams and staff's familiarity with them.

Item	*N*	Yes N(%)	No N(%)
Does your university organize examinations on basic medical knowledge?	1746	949 (54.4)	797 (45.6)
Does your university organize an examination on medical practice before students graduate?	1760	1489 (84.6)	271 (15.4)
Are you familiar with the Standard Patient (SP) exam?	1775	973 (54.8)	802 (45.2)
Does your university organize Standard Patient (SP) exams?	1644	754 (45.9)	890 (54.1)
Are you familiar with computer-based case simulation (CCS) examination?	1776	895 (50.4)	881 (49.6)
Does your university use computer-based case simulation (CCS) examinations?	1637	636 (38.9)	1001 (61.1)
Does your university analyze exam results centrally and send them back to relevant departments afterwards?	1749	1550 (88.6)	199 (11.4)
Are you familiar with portfolio assessment?	1759	528 (30.0)	1231 (70.0)

Influence of Gender

With regard to the exam on basic medical knowledge, both male and female groups reported more than 50% on *'Yes'* option, and there was no gender difference. As to the exam on medical practice before students graduate, both male and female groups reported more than 80% on *'Yes'* option, and there was no gender difference. For the familiarity of Standard Patient (SP) exam, male group (62.3%) reported significantly higher percentage than female group (48.9%). To the organization of Standard Patient (SP) exam, male group also (50.0%) reported significantly higher percentage than female group (43.0%). For the familiarity of computer-based case simulation (CCS) exam, there was significant gender difference, with male group (54.5%) reporting higher than female group (47.7%). With regard to the use of computer-based case simulation (CCS) examinations, both male and female groups reported less than 50%, and male group reported higher than female group. As to the analysis and feedback of exam results, both male and female groups reported higher than 80% on the *'Yes'* option, and there was no difference between these two gender groups. For the familiarity of portfolio assessment, both male and female groups reported less than 40% on *'Yes'* option, and male reported significantly higher than female group (see Table 6.29).

Table 6.29. Universities' exams and staff's familiarity about them by gender.

Item	Male N(%)		Female N(%)		χ^2	*P*
Does your university organize examinations on basic medical knowledge?	421	(54.0)	510	(55.1)	0.18	0.670
Does your university organize an examination on medical practice before students graduate?	648	(83.1)	810	(86.3)	3.36	0.067
Are you familiar with the Standard Patient (SP) exam?	490	(62.3)	463	(48.9)	31.04	<0.001
Does your university organize Standard Patient (SP) exams?	365	(50.0)	377	(43.0)	7.88	0.005
Are you familiar with computer-based case simulation (CCS) examination?	428	(54.5)	453	(47.7)	7.88	0.005
Does your university use computer-based case simulation (CCS) examinations?	309	(42.2)	316	(36.4)	5.61	0.018
Does your university analyze exam results centrally and send them back to relevant departments afterwards?	689	(88.6)	829	(89.0)	0.10	0.752
Are you familiar with portfolio assessment?	262	(33.5)	251	(26.8)	9.10	0.003

Note: The percentages presented in this table are percentage of option 'Yes' and same in the following tables unless it is indicated.

Influence of Age

For the exam on basic medical knowledge, all these four age groups reported higher than 50% on *'Yes'* option, and there was no difference among these age groups. For the exam on medical practice before students graduate, all the age groups reported higher than 80% on *'Yes'* option, and the age effect was not significant. With regard to the familiarity of Standard Patient (SP) exam, there was significant age effect, with Group above 43 reporting highest (59.8%) and Group 19–30 reporting lowest (45.9%). To the organization of Standard Patient (SP) exam, all these four age groups reported less than 50%, and there was no significant difference among the four groups. For the familiarity of computer-based case simulation (CCS) exam, all these age groups reported around 50% on it, and there was no age effect. With regard to the use of computer-based case simulation (CCS) examinations, all the age groups reported less than 50%, and there was no age effect. For the analysis and feedback of exam results, all these four age groups reported higher than 80%, and there was no significant difference among them. As to the familiarity of portfolio assessment, all the age groups reported less than 40%. The age effect was significant, with Group above 43 reporting highest and Group 31–35 reporting lowest (see Table 6.30).

Influence of Profession

For the exam on basic medical knowledge, there was significant profession effect, with Clinical teaching staff reporting highest (68.3%) and Public health teaching staff reporting lowest (38.7%). With regard to the exam on medical practice before students graduate, all the four profession groups reported more than 70% on *'Yes'* option. There was significant difference among these four profession groups, with Clinical teaching staff reporting highest and Public health teaching staff reporting lowest. To the familiarity of Standard Patient (SP) exam, there was significant profession effect, with Teaching administrators reporting highest (75.4%) and Public health teaching staff reporting lowest (35.5%). Concerning the organization of Standard Patient (SP) exam, there was significant profession effect, with Teaching administrators reporting highest (56.6%) and Public health teaching staff reporting lowest (26.4%). For the familiarity of computer-based case simulation (CCS) exam, the profession effect was significant, with Teaching administrators reporting highest (69.3%)

Table 6.30. Universities' exams and staff's familiarity about them by age.

Item	19–30 N(%)		31–35 N(%)		36–43 N(%)		>43 N(%)		χ^2	P
Does your university organize examinations on basic medical knowledge?	236	(58.7)	219	(52.1)	219	(53.5)	215	(51.9)	4.91	0.178
Does your university organize an examination on medical practice before students graduate?	351	(87.1)	371	(87.1)	348	(82.7)	336	(82.0)	7.40	0.060
Are you familiar with the Standard Patient (SP) exam?	186	(45.9)	228	(53.6)	249	(58.9)	251	(59.8)	20.07	<0.001
Does your university organize Standard Patient (SP) exams?	181	(46.6)	175	(44.3)	183	(47.5)	176	(46.2)	0.88	0.831
Are you familiar with computer-based case simulation (CCS) examination?	196	(48.2)	199	(46.8)	217	(51.1)	226	(54.1)	5.22	0.156
Does your university use computer-based case simulation (CCS) examinations?	152	(39.2)	141	(35.6)	149	(39.2)	157	(41.3)	2.76	0.429
Does your university analyze exam results centrally and send them back to relevant departments afterwards?	349	(87.0)	372	(88.6)	373	(89.9)	365	(88.4)	1.63	0.653
Are you familiar with portfolio assessment?	106	(26.0)	102	(24.4)	130	(31.1)	141	(34.1)	12.08	0.007

and Public health teaching staff reporting lowest (29.0%). With regard to the use of computer-based case simulation (CCS) examinations, all these four profession groups reported less than 50%. The profession effect was significant, Clinical teaching staff reported highest and Public health teaching staff reported lowest. For the analysis and feedback of exam results, all four profession groups reported more than 80% on '*Yes*', and there was no profession effect. As the familiarity of portfolio assessment, all profession groups reported less than 40%, there was significant profession effect, with Clinical teaching staff reporting highest and Public health teaching staff reporting lowest (see details in Table 6.31).

Table 6.31. Universities' exams and staff's familiarity about them by profession.

Item	Teaching administrators N(%)		Basic studies teaching staff N(%)		Public health teaching staff N(%)		Clinical teaching staff N(%)		χ^2	*P*
Does your university organize examinations on basic medical knowledge?	59	(46.5)	337	(47.0)	82	(38.7)	471	(68.3)	93.60	<0.001
Does your university organize an examination on medical practice before students graduate?	106	(83.5)	588	(82.2)	157	(73.0)	638	(90.8)	45.75	<0.001
Are you familiar with the Standard Patient (SP) exam?	95	(75.4)	293	(40.5)	78	(35.5)	507	(71.9)	198.23	<0.001
Does your university organize Standard Patient (SP) exams?	69	(56.6)	264	(41.6)	53	(26.4)	368	(53.6)	57.36	<0.001
Are you familiar with computer-based case simulation (CCS) examination?	88	(69.3)	289	(39.9)	63	(29.0)	455	(64.4)	145.06	<0.001
Does your university use computer-based case simulation (CCS) examinations?	56	(45.9)	211	(33.0)	46	(24.0)	323	(47.2)	49.80	<0.001
Does your university analyze exam results centrally and send them back to relevant departments afterwards?	108	(86.4)	634	(89.2)	194	(89.4)	614	(88.2)	1.07	0.785
Are you familiar with portfolio assessment?	42	(33.1)	176	(24.5)	40	(18.2)	270	(38.8)	51.25	<0.001

Influence of Job Title

For the exam on basic medical knowledge, all these three job title groups reported higher than 50% on *'Yes'* option. The significant job title effect indicated that Junior group reported highest and Middle group reported lowest. For the exam on medical practice before students graduate, all the job title groups reported higher than 80% on *'Yes'* option, and the job title effect was not significant. With regard to the familiarity of Standard Patient (SP) exam, all these three groups reported more than 50% on *'Yes'* option. There was significant job title effect, with Senior group reporting highest and Middle group reporting lowest. To the organization of Standard Patient (SP) exam, all these three job title groups reported less than 50%, and there was no significant difference among the three groups. For the familiarity of computer-based case simulation (CCS) exam, all these job title groups reported around 50% on it, and there was no job title effect. With regard to the use of computer-based case simulation (CCS) examinations, all the groups reported less than 50%, and there was no difference among these three job title groups. For the analysis and feedback of exam results, all these three job title groups reported higher than 80%, and there was no significant difference among them. As to the familiarity of portfolio assessment, all the job title groups reported less than 40%. The significant job title effect showed that the Middle group reported lowest (see Table 6.32).

Table 6.32. Universities' exams and staff's familiarity about them by job title.

Item	Junior N(%)		Middle N(%)		Senior N(%)		χ^2	*P*
Does your university organize examinations on basic medical knowledge?	259	(61.1)	298	(50.6)	392	(53.5)	11.33	0.003
Does your university organize an examination on medical practice before students graduate?	369	(86.2)	504	(84.6)	616	(83.7)	1.32	0.517
Are you familiar with the Standard Patient (SP) exam?	224	(52.3)	300	(50.0)	449	(60.1)	15.13	0.001
Does your university organize Standard Patient (SP) exams?	181	(45.1)	251	(45.0)	322	(47.0)	0.62	0.733
Are you familiar with computer-based case simulation (CCS) examination?	207	(48.1)	294	(49.0)	394	(52.8)	3.09	0.213
Does your university use computer-based case simulation (CCS) examinations?	171	(42.6)	205	(37.0)	260	(38.1)	3.38	0.185
Does your university analyze exam results centrally and send them back to relevant departments afterwards?	369	(87.0)	515	(87.4)	666	(90.5)	4.43	0.109
Are you familiar with portfolio assessment?	142	(33.2)	143	(24.1)	243	(33.0)	15.09	0.001

Influence of University's Geographical Location

With regard to the exam on basic medical knowledge, both East and West groups reported more than 50% on *'Yes'* option, and there was difference between East and West groups. As to the exam on medical practice before students graduate, both East and West groups reported more than 70% on '*Yes*' option, and East group reported significantly higher percentage than West group. For the familiarity of Standard Patient (SP) exam, East group (56.5%) reported significantly higher percentage than West (47.7%) did. To the organization of Standard Patient (SP) exam, both East and West groups reported less than 50% on '*Yes*' option, and East group reported significantly higher percentage than West group did. For the familiarity of computer-based case simulation (CCS) exam, East and West groups reported around 50% on '*Yes*' option and there was no significant difference between them. With regard to the use of computer-based case simulation (CCS) examinations, both East and West groups reported less than 50%, and East group reported higher than West group. As to the analysis and feedback of exam results, both East and West groups reported higher than 80% on the '*Yes*' option, and East group reported significantly higher than West group. For the familiarity of portfolio assessment, both East and West group reported less than 40% on '*Yes*' option, and East group reported significantly higher than West group (see Table 6.33).

Table 6.33. Universities' exams and staff's familiarity about them by university's geographical location.

Item	East N(%)		West N(%)		χ^2	*P*
Does your university organize examinations on basic medical knowledge?	775	(55.1)	174	(51.3)	1.55	0.213
Does your university organize an examination on medical practice before students graduate?	1216	(85.9)	273	(79.1)	9.86	0.002
Are you familiar with the Standard Patient (SP) exam?	809	(56.5)	164	(47.7)	8.79	0.003
Does your university organize Standard Patient (SP) exams?	638	(48.0)	116	(36.8)	12.82	<0.001
Are you familiar with computer-based case simulation (CCS) examination?	735	(51.3)	160	(46.6)	2.39	0.122
Does your university use computer-based case simulation (CCS) examinations?	526	(40.1)	110	(34.0)	4.08	0.043
Does your university analyze exam results centrally and send them back to relevant departments afterwards?	1269	(89.9)	281	(83.4)	11.36	0.001
Are you familiar with portfolio assessment?	457	(32.2)	71	(21.0)	16.17	<0.001

6.6 Summary

In order to understand staff's opinions on assessment in medicine and health education, this chapter was organized in five sections. First, general ideas about assessment were presented on the aspects of (1) the extent they agree that teaching evaluations are an important part of the teaching and learning process, (2) the extent they think that there should be an assessment of basic medical knowledge for students who have taken the relevant courses after their second and third years of study, (3) the extent they believe that there should be an assessment of medical practice before students graduate, (4) the extent they believe that the Standard Patient (SP) exam should be used on the medical practice assessment for graduation, (5) the extent they think that exam results should be analyzed centrally and sent back to relevant departments afterwards. Second, the methods should be used to evaluate teaching. Third, evaluation methods have been used at participants' universities. Fourth, an overview picture of the computer-based medical test database in participants' universities is presented. Lastly, the overall situation about medical universities' exams and staff's familiarity about it has been presented. Data analyses in this chapter do not only include overall findings to the above addressed issues, but also take into consideration the impact of gender, age, profession, job title, and the university's geographical location, respectively.

Sixteen questions were asked corresponding with the above mentioned aspects. There were five 5-point Likert-type questions, two multiple-choice questions, one single-choice question, and eight *'Yes/No'* questions. Overall findings to these questions were reported in the five sections in this chapter, and variables such as gender, age, profession, job title, and the university's geographical location were respectively taken into consideration of the analysis.

Overall findings in this chapter can be summarized in the following:

1. In general, participants, despite differences in their gender, age, profession, job title, university's geographical location, showed positive attitude towards the importance of teaching evaluations in the teaching and learning process, the assessment of basic medical knowledge, the assessment of medical practice before students graduate, the use of Standard Patient (SP) exam, and the analysis

and feedback of exam results. In particular, younger participants had a tendency of reporting a higher degree of importance level of teaching evaluations than their older peers. Teaching administrators reported a higher degree of importance level of concerning the majority of the items than other groups.

2. In spite of their agreement on the importance of teaching evaluations in the teaching and learning process, a strong disagreement among participants was found with regard to how to evaluate their teaching process. Concerning the methods which should be used to evaluate teaching, less than 50% participants agreed on Evaluation by management staff. In particular, a significantly higher level of disagreement could be found by female than male participants, younger than elder participants, Teaching administrators than other groups of teaching staff, Junior than Senior groups, and West than East groups. Agreement levels in relation to other options varied, nevertheless, most participants agreed on Evaluation by peers, Evaluation by one's self and Evaluation by students. The ranking of these four options was: Evaluation by peers, Evaluation by students, Evaluation by one's self, and Evaluation by management staff.
3. With regard to what kinds of evaluation methods have been used at the participants' universities, there were more than 50% participants reporting their experiences of Evaluation by management staff, Evaluation by peers, and Evaluation by students. Less than 50% participants reported that there was Evaluation by selves in their universities. The ranking of these four options was: Evaluation by students, Evaluation by peers, Evaluation by management staff, and Evaluation by selves.
4. Computer-based medical test database for student assessments were used in the majority of these investigated universities. Significant differences could be found by participants from different profession groups, with a tendency of more Clinical teaching staff and Teaching administrators responding *'Yes'* than the other groups, and university's geographical groups, with a tendency of more participants from East group responding*'Yes'* than West group. Nevertheless, given the context that computer-based medical test

database were introduced in all of these 22 universities at different levels, a considerably good number of participants responded *'No'* (15%), and *'Do not know'* (13.9%). These negative responses were scattered among different groups.

5. On participants' options on the usage and familiarity of new exams, there were more than 50% participants answering *'Yes'* on the exam on basic medical knowledge, the exam on medical practice before students graduate, the familiarity of Standard Patient (SP) exam, the familiarity of computer-based case simulation (CCS) exam, and the analysis and feedback of exam results. There were less than 50% participants answering *'Yes'* on the organization of Standard Patient (SP) exam, the use of computer-based case simulation (CCS) examinations, and the familiarity of portfolio assessment. The findings also indicated different levels of familiarity among the participants due to gender, age, profession, university geographical location. In general, male group, elder group, Teaching administrators, and East group seemed to have more familiarity to these new assessment methods than their peer groups.

To conclude, a strong agreement on the importance of teaching evaluations was reached by the participants in this study. They also reported certain level of knowledge and skills of the new assessment forms listed in this chapter. However, gender, age, profession, job title, university's geographical location made a difference in their access to the knowledge and skills in these new assessment methods. This indicates a thorough and comprehensive facilitation to a broader group of staff members to be familiar with new knowledge and methods in the educational reform.

7

Teaching Staff's Access to Educational Technologies

Xiaoju Duan, Jiannong Shi and Xiangyun Du

Abstract

This chapter aims to investigate institutional physical facilities for the support of educational reform from teaching staff's perspective. Four aspects are involved: (1) staff's satisfaction with physical facilities, (2) staff's access to educational technologies, (3) expected educational technology training, (4) wireless network on campus. Eight questions were asked in the survey. Data were analyzed and findings are presented in this chapter, which are structured in four sections. Section 1 reports participants' responses on (1) their satisfaction with the wireless network on campus, (2) the extent they think that the computer-based multimedia classrooms can meet the needs of teaching, (3) the extent they think that their universities provide sufficient online teaching resources, (4) the extent they think that there are well-functioning distance education systems and resources at their universities, and (5) the extent they are satisfied with the database retrieval system at their universities. Section 2 reports participants' choices on which educational technologies are accessible in their teaching process at their universities. Section 3 presents participants'

Change and Reform in Medicine and Health Education in China, 117–139.

expectations on educational technology training provided by their institutions. Section 4 reports participants' access to wireless network on campus. Data analyses in this chapter do not only include overall findings to the above addressed issues, but also take into consideration the impact of gender, age, profession, job title, and the university's geographical location, respectively.

7.1 General Opinions on Physical Facilities

In order to understand staff's overall satisfaction with the institutional physical facilities, five questions were asked that require 5-point Likert-type answers. These questions include: (1) Satisfaction with the wireless network on campus (see survey Q19 Appendix 1), (2) The extent they think that computer-based multimedia classrooms can meet the needs of teaching (see survey Q21 Appendix 1), (3) The extent they think that their university provides sufficient online teaching resources (see survey Q22 Appendix 1), (4) The extent they think that there are well-functioning distance education systems and resources at their universities (see survey Q23 Appendix 1), and (5) The extent they are satisfied with the database retrieval system at their university (see survey Q25 Appendix 1).

Overall findings

The overall descriptive analysis results of participants' responses of their satisfaction with the campus wireless network, the satisfaction of the computer-based multimedia classrooms, the sufficiency of university's online teaching resources, the sufficiency of university's distance education systems and resources, and the satisfaction of the university's database retrieval system, are shown in Table 7.1. In general, participants evaluated higher than the mean score on all these five questions.

Influence of Gender

Both male and female groups held positive attitudes to their university's physical facilities. Male and female groups had similar opinions on all these five aspects, namely, the satisfaction with the campus wireless network, the satisfaction of the computer-based multimedia classrooms, the sufficiency of university's online teaching resources, the sufficiency of university's distance

Table 7.1. General opinions on physical facility.

	N	*M*	*SD*
Are you satisfied with the wireless network on campus?	1801	3.49	1.05
To what extent do you think that computer-based multimedia classrooms can meet the needs of teaching?	1785	4.07	0.88
To what extent do you think that your university provides sufficient online teaching resources?	1777	3.67	1.02
To what extent do you think that there are well-functioning distance education systems and resources at your university?	1718	3.50	1.06
To what extent are you satisfied with the database retrieval system at your university?	1565	3.64	0.97

education systems and resources, the satisfaction of the university's database retrieval system (see Table 7.2).

Table 7.2. General opinions on physical facility by gender.

Item	Gender	*N*	*M*	*SD*	*t*	*P*
Are you satisfied with the wireless network on campus?	Male	801	3.49	1.10	−0.16	0.870
	Female	959	3.50	1.00		
To what extent do you think that computer-based multimedia classrooms can meet the needs of teaching?	Male	789	4.08	0.89	0.06	0.955
	Female	955	4.08	0.86		
To what extent do you think that your university provides sufficient online teaching resources?	Male	788	3.67	1.05	−0.13	0.896
	Female	949	3.68	0.99		
To what extent do you think that there are well-functioning distance education systems and resources at your university?	Male	759	3.52	1.09	0.79	0.428
	Female	919	3.48	1.03		
To what extent are you satisfied with the database retrieval system at your university?	Male	681	3.64	1.01	0.10	0.917
	Female	848	3.64	0.93		

Influence of Age

All these four age groups were positive to their university's physical facilities. Participants of different age groups had similar opinions on the satisfaction with the campus wireless network, and the sufficiency of university's distance education systems and resources. Participants of different age groups had different opinions on the satisfaction of the computer-based multimedia classrooms, the sufficiency of university's online teaching resources, and the satisfaction of the university's database retrieval system (see Table 7.3).

The *P*-values of post hoc tests between any two age groups are shown in Table 7.4. For the satisfaction of the computer-based multimedia classrooms,

Table 7.3. General opinions on physical facility by age.

Item	Age group	*N*	*M*	*SD*	*F*	*P*
Are you satisfied with the wireless network on campus?	19–30	410	3.43	1.04	2.12	0.096
	31–35	428	3.43	1.08		
	36–43	430	3.50	1.02		
	>43	431	3.58	1.04		
To what extent do you think that computer-based multimedia classrooms can meet the needs of teaching?	19–30	409	4.04	0.86	3.68	0.012
	31–35	425	4.03	0.91		
	36–43	425	4.04	0.89		
	>43	425	4.20	0.79		
To what extent do you think that your university provides sufficient online teaching resources?	19–30	408	3.65	1.01	4.13	0.006
	31–35	423	3.56	1.08		
	36–43	422	3.64	1.00		
	>43	426	3.80	0.98		
To what extent do you think that there are well-functioning distance education systems and resources at your university?	19–30	398	3.49	1.00	1.33	0.262
	31–35	410	3.43	1.08		
	36–43	410	3.48	1.05		
	>43	403	3.57	1.08		
To what extent are you satisfied with the database retrieval system at your university?	19–30	381	3.50	1.04	6.53	<0.001
	31–35	377	3.57	0.93		
	36–43	357	3.68	0.94		
	>43	361	3.79	0.91		

Table 7.4. General opinions on physical facility by age-post hoc analysis.

Item	19–30 vs 31–35	19–30 vs 36–43	19–30 vs >43	31–35 vs 36–43	31–35 vs >43	36–43 vs >43
Are you satisfied with the wireless network on campus?	0.982	0.295	0.034	0.280	0.030	0.275
To what extent do you think that computer-based multimedia classrooms can meet the needs of teaching?	0.855	0.948	0.009	0.905	0.005	0.007
To what extent do you think that your university provides sufficient online teaching resources?	0.174	0.863	0.039	0.231	0.001	0.024
To what extent do you think that there are well-functioning distance education systems and resources at your university?	0.358	0.871	0.293	0.446	0.048	0.222
To what extent are you satisfied with the database retrieval system at your university?	0.361	0.011	<0.001	0.102	0.002	0.129

Group above 43 evaluated higher than other three age groups. For the sufficiency of university's online teaching resources, Group above 43 also evaluated higher than other three age groups. For the satisfaction of the university's database retrieval system, Group above 43 evaluated higher than Group 19–30 and Group 31–35, and Group 36–43 evaluated higher than Group 19–30.

Influence of Profession

The four profession groups reported positive towards their university's physical facilities. Participants of different professions had different opinions on all these five questions, namely, their satisfaction with the campus wireless network, their satisfaction with the computer-based multimedia classrooms, the sufficiency of university's online teaching resources, the sufficiency of university's distance education systems and resources, and their satisfaction with the university's database retrieval system (see Table 7.5).

The P-values of post hoc tests between any two profession groups are shown in Table 7.6. For their satisfaction with the campus wireless network,

Table 7.5. General opinions on physical facilities by profession.

Item	Profession	*N*	*M*	*SD*	*F*	*P*
Are you satisfied with the wireless network on campus?	Teaching administrators	128	3.54	1.03	5.06	0.002
	Basic studies teaching staff	742	3.45	1.10		
	Public health teaching staff	219	3.29	1.05		
	Clinical teaching staff	712	3.59	0.99		
To what extent do you think that computer-based multimedia classrooms can meet the needs of teaching?	Teaching administrators	127	3.96	1.00	3.31	0.020
	Basic studies teaching staff	732	4.10	0.86		
	Public health teaching staff	221	3.93	0.90		
	Clinical teaching staff	705	4.11	0.85		
To what extent do you think that your university provides sufficient online teaching resources?	Teaching administrators	128	3.63	0.97	4.61	0.003
	Basic studies teaching staff	724	3.63	1.04		
	Public health teaching staff	220	3.50	1.06		
	Clinical teaching staff	705	3.77	1.00		
To what extent do you think that there are well-functioning distance education systems and resources at your university?	Teaching administrators	123	3.45	1.04	7.99	<0.001
	Basic studies teaching staff	698	3.46	1.05		
	Public health teaching staff	211	3.25	1.03		
	Clinical teaching staff	686	3.63	1.06		
To what extent are you satisfied with the database retrieval system at your university?	Teaching administrators	117	3.85	0.85	4.31	0.005
	Basic studies teaching staff	647	3.58	0.98		
	Public health teaching staff	197	3.54	0.97		
	Clinical teaching staff	604	3.71	0.96		

Table 7.6. General opinions on physical facility by profession-post hoc analysis.

Item	P1 v P2	P1 vs P3	P1 vs P4	P2 vs P3	P2 vs P4	P3 vs P4
Are you satisfied with the wireless network on campus?	0.374	0.034	0.642	0.050	0.013	<0.001
To what extent do you think that computer-based multimedia classrooms can meet the needs of teaching?	0.095	0.770	0.075	0.012	0.836	0.008
To what extent do you think that your university provides sufficient online teaching resources?	0.991	0.258	0.157	0.100	0.011	0.001
To what extent do you think that there are well-functioning distance education systems and resources at your university?	0.912	0.100	0.074	0.012	0.002	<0.001
To what extent are you satisfied with the database retrieval system at your university?	0.005	0.007	0.148	0.670	0.018	0.040

Note: P1, Teaching administrators; P2, Basic studies teaching staff; P3, Public health teaching staff, P4, Clinical teaching staff.

Teaching administrators, Basic studies teaching staff and Clinical teaching staff evaluated higher than public health teaching staff. For their satisfaction with the computer-based multimedia classrooms, Basic studies teaching staff and Clinical teaching staff evaluated higher than Public health teaching staff. For the sufficiency of university's online teaching resources, Clinical teaching staff evaluated higher than Basic studies teaching staff and Public health teaching staff. With regard to the sufficiency of university's distance education systems and resources, Clinical teaching staff evaluated higher than Basic studies teaching staff, which in turn evaluated higher than Public health teaching staff. As to the satisfaction with the university's database retrieval system, Teaching administrators and Clinical teaching staff evaluated higher than Basic studies teaching staff and Public health teaching staff.

Influence of Job Title

All the job title groups reported positive attitudes to their university's physical facilities. Participants of different job titles had different opinions on all the five questions: their satisfaction with the campus wireless network, their satisfaction with the computer-based multimedia classrooms, the sufficiency of university's online teaching resources, the sufficiency of university's distance education systems and resources, and their satisfaction with the university's database retrieval system (see Table 7.7).

The P-values of post hoc tests between any two title groups are shown in Table 7.8. With regard to the satisfaction with the campus wireless network, the Senior group evaluated higher than Junior and Middle groups. As to the satisfaction with the computer-based multimedia classrooms, Senior group also reported higher than Junior and Middle groups. For the sufficiency of university's online teaching resources, Junior and Senior groups evaluated

Table 7.7. General opinions on physical facility by job title.

Item	Job title	*N*	*M*	*SD*	*F*	*P*
Are you satisfied with the wireless network on campus?	Junior	436	3.44	1.08	3.26	0.039
	Middle	611	3.44	1.04		
	Senior	754	3.56	1.03		
To what extent do you think that computer-based multimedia classrooms can meet the needs of teaching?	Junior	432	4.03	0.87	4.78	0.009
	Middle	605	4.01	0.92		
	Senior	748	4.15	0.83		
To what extent do you think that your university provides sufficient online teaching resources?	Junior	430	3.71	1.02	7.32	0.001
	Middle	598	3.55	1.07		
	Senior	749	3.75	0.98		
To what extent do you think that there are well-functioning distance education systems and resources at your university?	Junior	419	3.57	1.04	5.61	0.004
	Middle	583	3.38	1.06		
	Senior	716	3.56	1.05		
To what extent are you satisfied with the database retrieval system at your university?	Junior	387	3.53	1.05	10.33	<0.001
	Middle	544	3.57	0.96		
	Senior	634	3.77	0.90		

Table 7.8. General opinions on physical facility by job title-post hoc analysis.

Item	Junior vs Middle	Junior vs Senior	Middle vs Senior
Are you satisfied with the wireless network on campus?	0.939	0.048	0.023
To what extent do you think that computer-based multimedia classrooms can meet the needs of teaching?	0.652	0.031	0.004
To what extent do you think that your university provides sufficient online teaching resources?	0.012	0.443	<0.001
To what extent do you think that there are well-functioning distance education systems and resources at your university?	0.006	0.885	0.003
To what extent are you satisfied with the database retrieval system at your university?	0.504	<0.001	<0.001

higher than Middle group. For the sufficiency of university's distance education systems and resources, Junior and Senior groups reported higher than Middle group. With regard to the satisfaction with the university's database retrieval system, Senior group evaluated higher than Junior and Middle group.

Influence of University's Geographical Location

Both East and West groups held positive attitudes towards their university's physical facilities. There were significant location effects. East group reported higher than West group on all these five questions: the satisfaction with the campus wireless network, the satisfaction with the computer-based multimedia classrooms, the sufficiency of university's online teaching resources, the sufficiency of university's distance education systems and resources, and the satisfaction with the university's database retrieval system (see Table 7.9).

Table 7.9. General opinions on physical facility by university geographical location.

Item	Location	*N*	*M*	*SD*	*t*	*P*
Are you satisfied with the wireless network on campus?	East	1450	3.58	1.02	7.74	<0.001
	West	351	3.11	1.06		
To what extent do you think that computer-based multimedia classrooms can meet the needs of teaching?	East	1437	4.11	0.85	3.82	<0.001
	West	348	3.91	0.95		
To what extent do you think that your university provides sufficient online teaching resources?	East	1430	3.73	1.02	5.02	<0.001
	West	347	3.43	1.00		
To what extent do you think that there are well-functioning distance education systems and resources at your university?	East	1389	3.56	1.05	4.84	<0.001
	West	329	3.25	1.04		
To what extent are you satisfied with the database retrieval system at your university?	East	1282	3.70	0.94	5.19	<0.001
	West	283	3.37	1.02		

7.2 Access of Educational Technologies

In order to obtain knowledge about staff's perception on the access to educational technologies, one multiple-choice question (see survey Q24 Appendix 1) was asked with six options to choose. The options were: (1) Projector and projection; (2) Computer-based multimedia; (3) TV and video; (4) Radio and recording; (5) Virtual reality; and (6) Online teaching.

Overall findings

Participants' reports on the access of the following educational technologies in the teaching process at their universities are shown in Table 7.10 ($N = 1781$). More than 60% participants chose the options of Projector and projection, Computer-based multimedia, and TV and video. Less than 40% participants chose options of Radio and recording, Virtual reality, and Online teaching. Results of participants responses on the six options can be ranked as following: Computer-based multimedia, Projector and projection, TV and video, Online teaching, Radio and recording, and Virtual reality.

Table 7.10. Accessible educational technologies.

Item	Yes N(%)		No N(%)	
Projector and projection	1360	(76.4)	421	(23.6)
Computer-based multimedia	1631	(91.6)	150	(8.4)
TV and video	1158	(65.1)	622	(34.9)
Radio and recording	557	(31.3)	1224	(68.7)
Virtual reality	486	(27.3)	1295	(72.7)
Online teaching	696	(39.1)	1085	(60.9)

Influence of Gender

More than 60% participants chose the options of Projector and projection, Computer-based multimedia, and TV and video in both male and female groups. Less than 50% participants chose the options of Radio and recording, Virtual reality, and Online teaching in both male and female groups. Male and female groups reported similar percentage on the options of Projector and projection, Computer-based multimedia, TV and video, and Online teaching. Male group reported higher than female group on the options of Radio and recording and Virtual reality. Both male and female groups had same rankings on the six options: Computer-based multimedia, Projector and projection, TV and video, Online teaching, Radio and recording, and Virtual reality (see Table 7.11).

Influence of Age

More than 60% participants chose the options of Projector and projection, Computer-based multimedia, and TV and video in all these four age groups.

Table 7.11. Accessible educational technologies by gender.

Item	Male N(%)		Female N(%)		χ^2	*P*
Projector and projection	591	(75.3)	733	(76.8)	0.51	0.475
Computer-based multimedia	721	(91.8)	875	(91.6)	0.03	0.866
TV and video	524	(66.8)	603	(63.2)	2.37	0.124
Radio and recording	273	(34.8)	271	(28.4)	8.21	0.004
Virtual reality	243	(31.0)	231	(24.2)	9.95	0.002
Online teaching	318	(40.5)	355	(37.2)	2.02	0.155

Less than 50% participants chose the options of Radio and recording, Virtual reality, and Online teaching in all these age groups. There was significant age effect on the option of Projector and projection, with Group 19–30 reporting the highest and Group above 43 reporting lowest. All these age groups had similar percentage on the options of Computer-based multimedia, TV and video, Radio and recording, Virtual reality, and Online teaching. All these four age groups had same rankings on these options: Computer-based multimedia, Projector and projection, TV and video, Online teaching, Radio and recording, and Virtual reality (see Table 7.12).

Table 7.12. Accessible educational technologies by age.

Item	19–30 N(%)		31–35 N(%)		36–43 N(%)		>43 N(%)		χ^2	*P*
Projector and projection	336	(82.8)	340	(79.8)	304	(71.7)	301	(70.8)	24.21	<0.001
Computer-based multimedia	364	(89.7)	392	(92.0)	385	(90.8)	401	(94.4)	6.69	0.083
TV and video	261	(64.3)	262	(61.6)	274	(64.6)	289	(68.0)	3.79	0.285
Radio and recording	132	(32.5)	138	(32.4)	127	(30.0)	130	(30.6)	0.97	0.809
Virtual reality	102	(25.1)	103	(24.2)	125	(29.5)	122	(28.7)	4.40	0.221
Online teaching	151	(37.2)	155	(36.4)	168	(39.6)	174	(40.9)	2.39	0.495

Influence of Profession

More than 60% participants chose the options of Projector and projection, Computer-based multimedia, and TV and video in all these four profession groups. There were less than 50% participants choosing the options of Radio and recording, and Virtual reality in all these profession groups. The percentage of choosing the option of Online teaching was between 34.4% to 52.0%.

There was a significant profession effect on the option of Projector and projection, with Clinical teaching staff reporting the highest and Basic studies teaching staff reporting the lowest. All these profession groups had similar high percentage on the option of Computer-based multimedia. For the option of TV and video, there was significant difference among these four profession groups, with Teaching administrators reporting highest and Basic studies teaching staff reporting lowest. For Radio and recording, the significant profession effect indicated that Teaching administrators reported the highest while basic studies reported lowest. Regarding the option of Virtual reality, the profession effect was significant, which showed that Clinical teaching staff reported highest and Public health teaching staff reported lowest. For the option of Online teaching, the significant profession effect indicated that Teaching administrators reported highest while Public health teaching staff reported lowest.

All these four profession groups had same rankings on these options: Computer-based multimedia, Projector and projection, TV and video, Online teaching, Radio and recording, and Virtual reality (see Table 7.13).

Influence of Job Title

There were more than 60% participants choosing the options of Projector and projection, Computer-based multimedia, and TV and video in all these three job title groups. There were less than 50% participants choosing the options of Radio and recording, Virtual reality, and Online teaching in all these job title groups.

There was significant job title effect on the option of Projector and projection, with Junior group reporting highest and Senior group reporting lowest. All these three job title groups had similar high percentage on the options of Computer-based multimedia and of TV and video. For the options of Radio and

Table 7.13. Accessible educational technologies by profession.

Item	Teaching administrators N(%)		Basic studies teaching staff N(%)		Public health teaching staff N(%)		Clinical teaching staff N(%)		χ^2	P
Projector and projection	100	(78.7)	500	(68.5)	175	(79.2)	585	(83.2)	44.70	<0.001
Computer-based multimedia	120	(94.5)	666	(91.2)	203	(91.9)	642	(91.3)	1.59	0.662
TV and video	90	(70.9)	439	(60.2)	134	(60.6)	495	(70.4)	20.16	<0.001
Radio and recording	56	(44.1)	163	(22.3)	58	(26.2)	280	(39.8)	63.43	<0.001
Virtual reality	46	(36.2)	148	(20.3)	33	(14.9)	259	(36.8)	72.55	<0.001
Online teaching	66	(52.0)	251	(34.4)	72	(32.6)	307	(43.7)	25.77	<0.001

recording, and Virtual reality, the three job title groups did not differ from each other. For the option of Online teaching, the significant job title effect indicated that Senior group reported highest while Middle group reported lowest.

All these three job title groups had same rankings on these options: Computer-based multimedia, Projector and projection, TV and video, Online teaching, Radio and recording, and Virtual reality (see details in Table 7.14).

Influence of University's Geographical Location

More than 50% participants chose the options of Projector and projection, Computer-based multimedia, and TV and video in both East and West groups. There were less than 50% participants choosing the options of Radio and recording, Virtual reality, and Online teaching in both East and West groups.

East and West groups reported similar percentage on the options of Projector and projection, and Computer-based multimedia. East group reported significant higher than West group on TV and video, Radio and recording, Virtual reality, and Online teaching.

Both East and West groups had same rankings on the six options: Computer-based multimedia, Projector and projection, TV and video, Online teaching, Radio and recording, and Virtual reality (see Table 7.15).

Table 7.14. Accessible educational technologies by job title.

Item	Junior N(%)		Middle N(%)		Senior N(%)		χ^2	P
Projector and projection	355	(82.8)	459	(75.9)	546	(73.1)	14.21	0.001
Computer-based multimedia	385	(89.7)	550	(90.9)	696	(93.2)	4.69	0.096
TV and video	278	(64.8)	382	(63.1)	498	(66.8)	1.94	0.380
Radio and recording	143	(33.3)	188	(31.1)	226	(30.3)	1.22	0.544
Virtual reality	126	(29.4)	145	(24.0)	215	(28.8)	5.14	0.077
Online teaching	176	(41.0)	210	(34.7)	310	(41.5)	7.37	0.025

Table 7.15. Accessible educational technologies by university's geographical location.

Item	East		West		χ^2	P
Projector and projection	1097	(76.5%)	263	(75.8%)	0.08	0.781
Computer-based multimedia	1317	(91.8%)	314	(90.5%)	0.66	0.416
TV and video	959	(66.9%)	199	(57.3%)	11.26	0.001
Radio and recording	483	(33.7%)	74	(21.3%)	19.85	<0.001
Virtual reality	413	(28.8%)	73	(21.0%)	8.49	0.004
Online teaching	606	(42.3%)	90	(25.9%)	31.27	<0.001

7.3 Expected Educational Technology Training

In order to understand staff's expectation on educational technology training provided by their universities, one multiple-choice question (see survey Q26 Appendix 1) was asked with eight options to choose. The options were: (1) Computer basis; (2) Office software; (3) Use of internet; (4) Database management; (5) PPT making; (6) Online teaching techniques; (7) Online teaching management; and (8) Online teaching technology.

Overall findings

Participants' overall expectations on educational technology training to be provided by their universities are shown in Table 7.16 ($N = 1760$). There were less than 50% participants choosing the options of Computer basis, Office software, Use of internet, Database management, Online teaching techniques, Online teaching management, and Online teaching technology. There were 61.5% participants choosing the option of PPT making. The overall ranking for these options was: PPT making, Database management, Online teaching techniques, Office software, Computer basis, Online teaching technology, Online teaching management, and Use of internet (Table 7.16).

Influence of Gender

Both male and female groups reported less than 50% on the options of Computer basis, Office software, Use of internet, Database management, Online teaching techniques, Online teaching management, and Online teaching technology. There were more than 50% participants choosing the option of PPT making in both male and female groups.

Table 7.16. Expected educational technology training.

Item	Yes N(%)		No N(%)	
Computer basis	769	(43.7)	991	(56.3)
Office software	805	(45.7)	955	(54.3)
Use of internet	574	(32.6)	1186	(67.4)
Database management	868	(49.3)	892	(50.7)
PPT making	1083	(61.5)	677	(38.5)
Online teaching techniques	819	(46.5)	941	(53.5)
Online teaching management	601	(34.2)	1158	(65.8)
Online teaching technology	653	(37.1)	1105	(62.9)

Male group had significantly higher percentage on the option of Computer basis than the female group. The difference between male and female groups on all the other seven options was not significant.

The overall ranking for these options was similar for male and female groups. The first most selected options were PPT making, Database management, and Online teaching techniques, and the least selected three options were Online teaching technology, Online teaching management, and Use of internet (see Table 7.17).

Table 7.17. Expected educational technology training by gender.

Item	Male N(%)		Female N(%)		χ^2	*P*
Computer basis	360	(46.3)	384	(40.8)	5.38	0.020
Office software	352	(45.3)	431	(45.8)	0.04	0.852
Use of internet	267	(34.4)	286	(30.4)	3.13	0.077
Database management	382	(49.2)	463	(49.2)	0.00	0.996
PPT making	492	(63.3)	561	(59.6)	2.55	0.111
Online teaching techniques	362	(46.6)	433	(46.0)	0.07	0.796
Online teaching management	278	(35.8)	305	(32.4)	2.26	0.133
Online teaching technology	279	(36.0)	354	(37.6)	0.46	0.500

Influence of Age

All the four age groups reported less than 50% on the options of Computer basis, Use of internet, Online teaching management, and Online teaching technology. There were more than 50% participants choosing the option of PPT making in all four age groups. The percentage on the options of Office software, Database management, and Online teaching techniques, varied around 50% in the age groups.

There was significant age effect on the option of PPT making, with Group above 43 reported highest and 19–30 reported lowest. The significant age effect indicated that Group 36–43 reporting highest and Group 19–30 reporting lowest on the option of Online teaching techniques. The difference among age groups on all the other six options was not significant.

The overall ranking for these options was slightly different in these four age groups. PPT making, Office software, and Database management were ranked as the highest three options in Group 19–30. PPT making, Database management, Online teaching techniques were ranked as the highest three options in other three age groups. For all these four age groups, the least selected three options were Online teaching technology, Online teaching management, and Use of internet (see Table 7.18).

Table 7.18. Expected educational technology training by age.

Item	19–30 N(%)		31–35 N(%)		36–43 N(%)		>43 N(%)		χ^2	P
Computer basis	187	(46.4)	166	(39.2)	76	(42.0)	173	(41.8)	4.46	0.216
Office software	203	(50.4)	178	(42.1)	182	(43.4)	183	(44.2)	6.68	0.083
Use of internet	119	(29.5)	124	(29.3)	133	(31.7)	150	(36.2)	5.92	0.116
Database management	196	(48.6)	197	(46.6)	209	(49.9)	210	(50.7)	1.64	0.650
PPT making	218	(54.1)	267	(63.1)	261	(62.3)	262	(63.3)	10.02	0.018
Online teaching techniques	159	(39.5)	186	(44.0)	216	(51.6)	207	(50.0)	15.45	0.001
Online teaching management	136	(33.7)	135	(31.9)	151	(36.0)	135	(32.7)	1.81	0.612
Online teaching technology	143	(35.5)	159	(37.6)	148	(35.5)	158	(38.2)	1.05	0.790

Influence of Profession

There was a significant profession effect on the option of Computer basis with Clinical teaching staff reporting the highest (54.3%) and Public health teaching staff reporting the lowest (29.2%). On the option of Office software, all the four profession groups reported lower than 50% and there was no profession effect. And, on the option of Use internet, all profession groups reported lower than 40%, and the significant profession effect was found that Clinical teaching staff reported the highest and Public health teaching staff reported the lowest. Regarding the option of Database management, there was a significant profession effect, with Clinical teaching staff reporting the highest (56.5%) and Basic studies teaching staff reporting the lowest (40.9%). More than 50% of all participants selected the option of PPT making. There were significant differences among these four profession groups with Clinical teaching staff reporting the highest and Teaching administrators reporting the lowest. Concerning the option of Online teaching techniques, there were significant profession differences, with Teaching administrators reporting the highest and Basic studies teaching staff reporting the lowest. On the option of Online teaching management, there were significant profession differences. Here, Teaching administrators reported the highest and Basic studies teaching staff reported the lowest. All these four profession groups reported lower than

50% on the option of Online teaching technology and there was no difference among the profession groups (see details in Table 7.19).

Table 7.19. Expected educational technology training by profession.

Item	Teaching administrators N(%)		Basic studies teaching staff N(%)		Public health teaching staff N(%)		Clinical teaching staff N(%)		χ^2	P
Computer basis	45	(35.4%)	283	(39.3)	63	(29.2)	378	(54.3)	59.72	<0.001
Office software	58	(45.7%)	318	(44.1)	89	(41.2)	340	(48.9)	5.28	0.152
Use of internet	37	(29.1%)	226	(31.3)	46	(21.3)	265	(38.1)	23.26	<0.001
Database management	65	(51.2%)	295	(40.9)	115	(53.2)	393	(56.5)	36.10	<0.001
PPT making	69	(54.3%)	412	(57.1)	133	(61.6)	469	(67.4)	18.72	<0.001
Online teaching techniques	65	(51.2%)	304	(42.2)	101	(46.8)	349	(50.1)	10.29	0.016
Online teaching management	74	(58.3%)	200	(27.8)	76	(35.2)	251	(36.1)	47.08	<0.001
Online teaching technology	56	(44.4%)	258	(35.8)	84	(39.1)	255	(36.6)	3.87	0.276

The overall rankings for these options were different in these four profession groups. The ranking by Teaching administrator was Online teaching management, PPT making, Database management and Online teaching techniques, Office software, Online teaching technology, Computer basis, and Use of internet. The ranking by Basic studies teaching staff was PPT making, Office software, Online teaching techniques, Database management, Computer basis, Online teaching technology, Use of internet, and Online teaching management. The ranking by Public health teaching staff was PPT making, Database management, Online teaching techniques, Office software, Online teaching technology, Online teaching management, Computer basis, and Use of internet. The ranking for Clinical teaching staff was PPT making, Database management, Computer basis, Online teaching techniques, Office software, Use of internet, Online teaching technology, and Online teaching management.

Influence of Job Title

All the three job title groups reported less than 50% on the options of Use of internet, Online teaching management, and Online teaching technology. There were more than 50% participants choosing the option of PPT making in all four age groups. The percentage of participants selecting the options of

Computer basis, Office software, Database management, and Online teaching techniques, varied from 38.1% to 53.8% in the job title groups.

There were significant job title effects on the five out of eight options, namely, Computer basis, Office software, Use of internet, Database management, and Online teaching techniques. Junior group reported highest on the options of Computer basis and Office software, while Senior group reported highest on the options of Use of internet and Online teaching techniques, and Middle group reported lowest on these five options. There was no difference among these five job title groups on the options of PPT making, Online teaching management, and Online teaching technology.

The overall rankings for these options were similar in three job title groups. PPT making was ranked as the most selected option, while Online teaching management and Use of internet were ranked as the least options in all job title groups (see Table 7.20).

Influence of University's Geographical Location

Both East and West groups reported less than 50% on the options of Computer basis, Office software, Use of internet, Database management, Online teaching techniques, Online teaching management, and Online teaching technology. There were more than 60% participants choosing the option of PPT making in both East and West groups.

East group had significantly higher percentage on the Use of internet and Online teaching techniques than West group. The differences between East and West groups on all the other options were not significant.

The overall rankings for these options were similar for East and West groups. The first two most selected options were PPT making and Database

Table 7.20. Expected educational technology training by job title.

Item	Junior N(%)		Middle N(%)		Senior N(%)		χ^2	*P*
Computer basis	231	(53.8)	228	(38.1)	310	(42.3)	26.23	<0.001
Office software	227	(52.9)	247	(41.2)	331	(45.2)	13.87	0.001
Use of internet	149	(34.7)	167	(27.9)	258	(35.2)	9.29	0.010
Database management	223	(52.0)	264	(44.1)	381	(52.0)	9.99	0.007
PPT making	256	(59.7)	357	(59.6)	470	(64.2)	3.79	0.151
Online teaching techniques	190	(44.3)	255	(42.6)	374	(51.1)	10.77	0.005
Online teaching management	149	(34.7)	212	(35.4)	240	(32.8)	1.04	0.594
Online teaching technology	151	(35.2)	228	(38.1)	274	(37.5)	0.96	0.619

management, and the least selected three options were Online teaching technology, Online teaching management, and Use of internet (see Table 7.21).

Table 7.21. Expected educational technology training by university's geographical location.

Item	East N(%)		West N(%)		χ^2	*P*
Computer basis	632	(44.6)	137	(40.1)	2.28	0.131
Office software	662	(46.7)	143	(41.8)	2.64	0.104
Use of internet	487	(34.3)	87	(25.4)	9.94	0.002
Database management	707	(49.9)	161	(47.1)	0.85	0.355
PPT making	872	(61.5)	211	(61.7)	0.01	0.945
Online teaching techniques	681	(48.0)	138	(40.4)	6.52	0.011
Online teaching management	494	(34.9)	107	(31.3)	1.57	0.211
Online teaching technology	534	(37.7)	119	(34.9)	0.92	0.339

7.4 Access to the Campus Wireless Network

In order to gain knowledge about staff's overall perception on the access to the campus wireless network, one question with two '*Yes/No*' sub-questions (see survey Q20 Appendix 1) were asked. The two sub-questions are: (1) Do you have access to the wireless network on campus? And (2) Is there campus wide wireless network in your university?

Overall findings

The descriptive analysis results of participants' options on the access to the on campus wireless network, and the availability of on campus wireless network is shown in Table 7.22. In general, there were 44% participants reporting that they had access to the campus wireless network while 72.7% participants reported that there was availability of campus wireless network in their universities.

Influence of Gender

Less than 50% participants reported that they had access to the campus wireless network in both male and female groups. There were more than 70%

Table 7.22. Access to the campus wireless network.

Item	*N*	Yes	No
Do you have the access to the wireless network on campus?	1820	800 (44.0%)	1020 (56.0%)
Is there an on campus wireless network in your university?	1820	1324 (72.7%)	496 (27.3%)

participants reporting availability of campus wide wireless network in their universities in both male and female groups. Male and female groups reported similar percentage regarding their access to the campus wireless network, and the availability of the campus wide wireless network (see Table 7.23).

Table 7.23. Access to the campus wireless network by gender.

Item	Male N(%)		Female N(%)		χ^2	P
Do you have the access to the campus wireless network?	335	(41.6)	442	(45.5)	2.74	0.098
Is there campus wireless network in your university?	580	(72.0)	706	(72.6)	0.10	0.752

Note: The percentages presented in this table are percentage of option 'Yes' and same in the following tables unless it is indicated.

Influence of Age

Less than 50% participants reported they had access to the campus wireless network in all four age groups. The significant age effect indicated that Group above 43 reported the highest and Group 19–30 reported the lowest. More than 60% participants reported that there was availability of campus wireless network in their universities in all age groups and the groups did not differ from each other (see Table 7.24).

Table 7.24. Access to the campus wireless network by age.

Item	19–30 N(%)		31–35 N(%)		36–43 N(%)		>43 N(%)		χ^2	P
Do you have the access to the campus wireless network?	162	(39.2)	170	(39.3)	195	(44.8)	212	(48.6)	11.09	0.011
Is there campus wireless network in your university?	297	(71.9)	300	(69.3)	321	(73.8)	318	(72.9)	2.47	0.481

Influence of Profession

Less than 50% participants reported that they had access to the campus wireless network in all four profession groups. The significant profession effect indicated that Clinical teaching staff reported the highest and Basic studies teaching staff reported the lowest. More than 60% participants reported that there was availability of campus wireless network in their universities in all profession groups. The profession effects were significant with Clinical teaching staff reporting highest and Basic studies teaching staff reporting lowest (see Table 7.25).

Table 7.25. Access to the campus wireless network by profession.

Item	Teaching administrators N(%)		Basic studies teaching staff N(%)		Public health teaching staff N(%)		Clinical teaching staff N(%)		χ^2	*P*
Do you have the access to the campus wireless network?	53	(41.4)	307	(40.9)	91	(41.2)	349	(48.5)	9.88	0.020
Is there campus wireless network in your university?	97	(73.4)	509	(67.8)	150	(67.9)	571	(79.3)	27.66	<0.001

Influence of Job Title

Generally, all the three job title groups reported less than 50% on their access to the campus wireless network. The significant job title effect showed that Senior group reported highest and Middle group reported lowest. There were more than 60% participants reporting the availability of campus wireless network in their universities in these job title groups and there was no significant difference among them (see details in Table 7.26).

Table 7.26. Access to the campus wireless network by job title.

Item	Junior N(%)		Middle N(%)		Senior N(%)		χ^2	*P*
Do you have the access to the campus wireless network?	193	(44.1)	236	(38.1)	371	(48.6)	15.29	<0.001
Is there campus wireless network in your university?	331	(75.6)	429	(69.3)	564	(73.9)	5.99	0.050

Influence of University's Geographical Location

There were less than 50% participants reporting their access to the campus wireless network in both East and West groups. There were more than 60% participants reporting the availability of campus wireless network in their universities in both East and West groups. The East group reported significantly higher than the West group on the access to the campus wireless network, and the availability of campus wireless network (see Table 7.27).

Table 7.27. Access to the campus wireless network by university's geographical location.

Item	East N(%)		West N(%)		χ^2	P
Do you have the access to the campus wireless network?	684	(46.8)	116	(32.3)	24.61	<0.001
Is there a campus wireless network in your university?	1086	(74.3)	238	(66.3)	9.39	0.002

7.5 Summary

In order to have staff's opinions on institutional physical facilities, this chapter consists of four sections. First, it deals with staff's satisfaction with physical facilities in their universities and about their satisfaction with the campus wireless network, the extent they thought that computer-based multimedia classrooms can meet the needs of teaching, the extent they thought that their university provides sufficient online teaching resources, the extent they thought that there are well-functioning distance education systems and resources, and the extent they were satisfied with the database retrieval system. Second, it intends to understand the participants' access to educational technology. Third, it investigates staff's expected educational technology training. Finally, it describes the access and availability of the campus wireless network in their universities.

Eight questions were employed in respective to the above mentioned aspects. There were five 5-point Likert-type questions, two multiple-choice questions and one question which consisted of two *'Yes/No'* sub-questions. Overall findings to these questions have been reported in the four sections in this chapter, and variables such as gender, age, profession, job title, and the university's geographical location were respectively taken into consideration of analysis.

Overall findings in this chapter can be summarized in the following:

1. In general, participants were positive on the satisfaction with the campus wireless network, the satisfaction with the computer-based multimedia classrooms, the sufficiency of university's online teaching resources, the sufficiency of university's distance education systems and resources, the satisfaction with their university's database retrieval system. Nevertheless, significant differences

were found among different profession groups and age groups. Middle group by job title showed considerably lower satisfaction level in most of the items than Junior and Senior groups. In addition, West group reported significantly lower satisfaction level than East group.

2. Concerning staff's access to the following educational technologies in the teaching process at their universities, the results showed a generally common access to the technologies such as Computer-based multimedia, Projector and projection, TV and video (over 50% respondents) in their teaching practice; however, much less access to Online teaching, Radio and recording, Virtual reality (less than 50% respondents). Although a general positive access to the basic teaching technologies such as Computer-based multimedia and Projector and projection, significant differences were identified due to profession, age, and job titles. For example, Clinical teaching staff reported highest and Basic studies teaching staff reported lowest. A decrease trend of access was identified by younger group than elder group and Junior group than Middle and Senior groups. University's geographical location does not make a difference in their staff's access to basic technology. However, East and West group differ significantly in the other four aspects, with the West group reporting much lower access to TV and video, Online teaching, Radio and recording, and Virtual reality.
3. With regard to the expected educational technology training, PPT making is the most selected option which had more than 50% participants choosing it. The rest of the seven options, Computer basis, Office software, Use of internet, Database management, Online teaching techniques, Online teaching management, and Online teaching technology, were agreed by less than 50% participants. Significant differences were found among profession groups and job title groups in the most of aspects.
4. For the staff's responses on the campus wireless network, there were less than 50% participants reporting they had access to the campus wireless network while there were more than 70% participants reporting that the campus wireless network was available in their universities. Significant differences were found among age

groups, profession groups, job title groups and between West and East groups. Elder group and Clinical teaching staff have considerably more access to the campus wireless network than their peer groups. Middle group and Basic studies teaching staff have considerably fewer access to the campus wireless network than their peer groups. West group reported significantly less access the on campus wireless network than East group.

In conclusion, participants in this study generally held positive attitude to the present campus physical facilities. Nevertheless, a few factors impact their access to the educational facilities, for example, profession, job title and university's geographical location. Teaching administrators and Clinical teaching staff have generally better access to the educational technologies than Basic studies teaching staff and Public health teaching staff. Middle group by job title has less access to the technologies than other groups. West group has much less access than East group. The majority of participants reported needs for educational technology training in certain aspects. These findings indicate that for the further work of reform, there is a general need for better distribution of educational resources by taking into the consideration of all these influence factors in this study. There is also a great demand for institutionalized support for all staff with better skills in using educational technologies.

8

Institutional Support for Teaching Staff's Participation in Educational Research and Reform

Xiaoju Duan, Xiangyun Du and Jiannong Shi

Abstract

The purpose of this chapter is to gain an overall knowledge about institutional support for teaching staff's participation in educational research and reform. Specifically, it investigates the participants' opinions on the necessity and current situation of establishing a unit focused on medicine and health education research, the involvement of educational research projects, as well as staff's publication on teaching and learning. Five questions were asked regarding these aspects in the survey. Data are analyzed and findings are presented in this chapter, which are structured in four sections. Section 1 reports how participants scaled to what extent they agreed with the necessity of establishing a unit focused on medicine and health education research. Section 2 shows participants' answers to whether there was such a unit in their university, and whether they had been involved in any educational research projects. Section 3 reports the levels of research projects they were involved if they answered '*Yes*' to the previous question. Section 4 reports the number of articles they

Change and Reform in Medicine and Health Education in China, 141–156.

published on medical teaching and learning research. Data analyses in this chapter not only include overall findings to the above addressed issues, but also take into consideration of the impact of gender, age, profession, job title, and the university's geographical locations respectively.

8.1 Opinion on the Necessity of a Research Unit

In order to understand staff's opinion on the necessity of establishing a unit focused on medicine and health education research, one question was asked that requires 5-point Likert-type answers. The question was "To what extent do they agree with the necessity of establishing a unit focused on medicine and health education research?" (see survey Q44 Appendix 1).

Overall findings

The descriptive data to this question is presented in Table 8.1. In general, participants positively responded to the necessity of establishing a unit focused on medicine and health education research.

Influence of Gender

Both male and female groups held positive attitude to the necessity of establishing a unit focused on medicine and health education research. Nevertheless, the significant gender effect showed that male group evaluated higher than female group (see Table 8.2).

Table 8.1. Opinion on the necessity of a research unit.

Item	*N*	*M*	*SD*
To what extent do you agree with the necessity of establishing a unit focused on medicine and health education research?	1771	4.29	0.74

Table 8.2. Opinion on the necessity of an research unit by gender.

Item	Gender	*N*	*M*	*SD*	*t*	*P*
To what extent do you agree with the necessity of establishing a unit focused on medicine and health education research?	Male	787	4.33	0.76	2.36	0.018
	Female	944	4.25	0.74		

Influence of Age

All the participants in these four age groups were positive to the necessity of establishing a unit focused on medicine and health education research. The age groups differed in the degree of agreement (see Table 8.3).

The P-values of post hoc tests between any two age groups are shown in Table 8.4. Group 19–30 and Group above 43 evaluated higher than Group 31–35 and Group 36–43 on the necessity of establishing a unit focused on medicine and health education research.

Influence of Profession

All the participants were positive to the necessity of establishing a unit focused on medicine and health education research in these four profession groups. Significant profession effect was identified on this question (see Table 8.5).

Table 8.3. Opinion on the necessity of a research unit by age.

Item	Age group	*N*	*M*	*SD*	*F*	*P*
To what extent do you agree with the necessity of establishing a unit focused on medicine and health education research?	19–30	405	4.34	0.77	2.90	0.034
	31–35	419	4.24	0.76		
	36–43	421	4.24	0.75		
	>43	426	4.35	0.69		

Table 8.4. Opinion on the necessity of a research unit by age-post hoc analysis.

Item	19–30 vs 31–35	19–30 vs 36–43	19–30 vs >43	31–35 vs 36–43	31–35 >43	36–43 vs >43
To what extent do you agree with the necessity of establishing a unit focused on medicine and health education research?	0.044	0.052	0.863	0.944	0.027	0.032

Table 8.5. Opinion on the necessity of a research unit by profession.

Item	Profession	*N*	*M*	*SD*	*F*	*P*
To what extent do you agree with the necessity of establishing a unit focused on medicine and health education research?	Teaching administrators	127	4.51	0.74	9.28	<0.001
	Basic studies teaching staff	721	4.22	0.77		
	Public health teaching staff	220	4.18	0.76		
	Clinical teaching staff	703	4.35	0.70		

The *P*-values of post hoc tests between any two profession groups are shown in Table 8.6. Concerning the necessity of establishing a unit focused on medicine and health education research, Teaching administrators evaluated higher than Clinical teaching staff, which in turn evaluated higher than Basic studies teaching staff and Public health teaching staff.

Influence of Job Title

Participants held positive attitude to the necessity of establishing a unit focused on medicine and health education research in all these three job title groups. The difference among these three job title groups was not significant on this question (see Table 8.7).

The *P*-values of post hoc tests between any two title groups are shown in Table 8.8. There was no difference between any two job title groups.

Influence of University's Geographical Location

Both East and West groups reported positively to the necessity of establishing a unit focused on medicine and health education research. East group reported significantly higher than West group (see Table 8.9).

Table 8.6. Opinion on the necessity of a research unit by profession-post hoc analysis.

Item	P1 vs P2	P1 vs P3	P1 vs P4	P2 vs P3	P2 vs P4	P3 vs P4
To what extent do you agree with the necessity of establishing a unit focused on medicine and health education research?	<0.001	<0.001	0.023	0.544	0.001	0.003

Note: P1, Teaching administrators; P2, Basic studies teaching staff; P3, Public health teaching staff, P4, Clinical teaching staff.

Table 8.7. Opinion on the necessity of a research unit by job title.

Item	Job title	*N*	*M*	*SD*	*F*	*P*
To what extent do you agree with the necessity of establishing a unit focused on medicine and health education research?	Junior	429	4.32	0.74	2.00	0.136
	Middle	594	4.24	0.77		
	Senior	748	4.30	0.72		

Table 8.8. Opinion on the necessity of a research unit by job title-post hoc analysis.

Item	Junior vs Middle	Junior vs Senior	Middle vs Senior
To what extent do you agree with the necessity of establishing a unit focused on medicine and health education research?	0.074	0.708	0.099

Table 8.9. Opinion on the necessity of an research unit by university geographical location.

Item	Location	*N*	*M*	*SD*	*t*	*P*
To what extent do you agree with the necessity of establishing a unit focused on medicine and health education research?	East	1425	4.30	0.73	2.02	0.044
	West	346	4.21	0.78		

8.2 Current Situation of Research Unit and Project Involvement

In order to obtain the knowledge about the current situation of medicine and health education research unit, and participants' involvement in educational research projects, two '*Yes/No*' questions were used. These questions are: (1) Whether their universities have a unit focused on medicine and health education research (see survey Q43 Appendix 1), and (2) Whether they have been or are currently involved in any educational research projects (see survey Q45 Appendix 1).

Overall findings

Generally speaking, there were 72.5% participants responding that their universities had a unit focused on medicine and health education research. Half of them had been involved in educational research projects (see Table 8.10).

Influence of Gender

Both male and female groups reported higher than 70% on the '*Yes*' answer of whether or not their universities had a unit focused on medicine and health education research. Furthermore, both groups reported similar percentage on it. Concerning the involvement in educational research project, male group (54.7%) reported higher than female group (46.5%) (see Table 8.11).

Table 8.10. Current situation of research unit and project involvement.

Item	*N*	Yes N(%)	No N(%)
Does your university have a unit focused on medicine and health education research?	1739	1261 (72.5)	478 (27.5)
Have you been or are you currently involved in any educational research projects?	1781	892 (50.1)	889 (49.9)

Table 8.11. Current situation of research unit and project involvement by gender.

Item	Male N(%)		Female N(%)		χ^2	*P*
Does your university have a unit focused on medicine and health education research?	569	(73.4)	669	(72.5)	0.19	0.665
Have you been or are you currently involved in any educational research projects?	431	(54.7)	443	(46.5)	11.49	0.001

Influence of Age

All these four age groups reported more than 60% on the existence of a unit focused on medicine and health education research in their universities. There was significant difference among these four age groups, with Group above 43 reporting highest and Group 31–35 reported lowest. With regard to the involvement in educational research, the age effect was significant. Group above 43 (63.1%) reported highest and Group 19–30 reported lowest (19–30%) (see Table 8.12).

Table 8.12. Current situation of research unit and project involvement by age.

Item	19–30 N(%)		31–35 N(%)		36–43 N(%)		>43 N(%)		χ^2	*P*
Does your university have a unit focused on medicine and health education research?	294	(73.0)	283	(69.0)	295	(71.8)	327	(78.6)	10.25	0.017
Have you been or are you currently involved in any educational research projects?	148	(36.2)	202	(47.5)	227	(53.8)	267	(63.1)	63.77	<0.001

Influence of Profession

All the four profession groups reported more than 70% on the existence of a unit focused on medicine and health education research in their universities. The significant profession effect showed that Teaching administrators evaluated highest and Clinical teaching staff evaluated lowest. On the involvement in educational research, the profession effect was significant, with Teaching administrators reporting highest and Clinical teaching staff reporting lowest (see details in Table 8.13).

Table 8.13. Current situation of research unit and project involvement by profession.

Item	Teaching administrators N(%)		Basic studies teaching staff N(%)		Public health teaching staff N(%)		Clinical teaching staff N(%)		χ^2	*P*
Does your university have a unit focused on medicine and health education research?	106	(82.8)	501	(71.5)	160	(74.1)	494	(71.2)	8.08	0.044
Have you been or are you currently involved in any educational research projects?	89	(69.5)	379	(51.8)	105	(47.9)	319	(45.4)	26.90	<0.001

Influence of Job Title

Participants from each job title group reported more than 60% on the existence of a unit focused on medicine and health education research in their universities. There was significant difference among these three job title groups, with Senior group reporting highest and Middle group reporting lowest. For the involvement in educational research, the significant job title effect indicated that Senior group reported highest while Junior group reported lowest (see details in Table 8.14).

Table 8.14. Current situation of research unit and project involvement by job title.

Item	Junior N(%)		Middle N(%)		Senior N(%)		χ^2	*P*
Does your university have a unit focused on medicine and health education research?	302	(71.2)	397	(68.8)	562	(76.2)	9.24	0.010
Have you been or are you currently involved in any educational research projects?	166	(38.4)	274	(45.5)	452	(60.5)	60.98	<0.001

Influence of University's Geographical Location

Both East and West groups reported higher than 60% on the existence of a unit focused on medicine and health education research in their universities. Furthermore, East group reported significantly higher than West group on it. Concerning the involvement in educational research project, there was no significant difference between East group (49.8%) and West group (51.3%) (see Table 8.15).

Table 8.15. Current situation of research unit and project involvement by university's geographical location.

Item	East N(%)		West N(%)		χ^2	*P*
Does your university have a unit focused on medicine and health education research?	1044	(74.6)	217	(64.0)	15.27	<0.001
Have you been or are you currently involved in any educational research projects?	714	(49.8)	178	(51.3)	0.25	0.615

8.3 Research Projects Level

In order to know detailed information on participants' involvement in educational research, one multiple-choice question (see survey Q46 Appendix 1) was asked with four options to choose. The options were: (1) National level; (2) Provincial level; (3) Municipal level; and (4) University level.

Overall findings

The results of participants' involvement in research projects are shown in Table 8.16 ($N = 946$). There were less than 30% participants reporting their involvement in National and Municipal level research projects. There were more than 50% participants reporting their involvement in Provincial and University level research projects.

Influence of Gender

There were less than 30% reporting on the levels of National and Municipal research projects in both male and female groups. There were more than 60% participants choosing University level research project in both male and female

Table 8.16. Levels of research projects.

Item	Yes N(%)		No N(%)	
National	168	(17.8)	777	(82.2)
Provincial	489	(51.8)	455	(48.2)
Municipal	234	(24.8)	710	(75.2)
University	621	(65.6)	325	(34.4)

groups. No significant difference was identified between male and female groups on the levels of National, Municipal and University research projects. Concerning the Provincial level research project, male group (56.1%) reported significantly higher percentage than female group (47.9%) (see Table 8.17).

Table 8.17. Levels of research projects by gender.

Item	Male N(%)		Female N(%)		χ^2	P
National	84	(18.6)	80	(16.9)	0.46	0.497
Provincial	253	(56.1)	227	(47.9)	6.24	0.013
Municipal	121	(26.8)	106	(22.4)	2.49	0.115
University	293	(64.8)	313	(65.9)	0.12	0.732

Influence of Age

For the option of National level, all these four age groups reported less than 30% and there was no difference among age groups. As to the option of Provincial level, four age groups reported around 50% and the age difference was not significant. Regarding the option of Municipal level, the age effect was significant with Group above 43 reported highest (25.2%) and Group 31–35 reported lowest. Concerning the option of University level, participants reported more than 50% in all these age groups. The significant age effect showed that Group above 43 reported the highest and Group 19–30 reported the lowest (see Table 8.18).

Table 8.18. Levels of research projects by age.

Item	19–30 N(%)		31–35 N(%)		36–43 N(%)		>43 N(%)		χ^2	P
National	35	(21.7)	36	(16.4)	46	(19.5)	42	(15.1)	3.89	0.274
Provincial	81	(50.3)	97	(44.3)	128	(54.2)	155	(55.8)	7.37	0.061
Municipal	36	(22.4)	41	(18.7)	72	(30.5)	70	(25.2)	9.03	0.029
University	84	(52.2)	149	(67.4)	153	(64.8)	195	(70.1)	15.41	0.002

Influence of Profession

For the option of National level, all these four profession groups reported less than 20% and there was no significant difference among profession groups. As to the option of Provincial level, four profession groups reported around 50% and the profession difference was not significant. Regarding the option of Municipal level, the profession effect was significant with Clinical teaching staff reporting highest (37.7%) and Teaching administrators reporting lowest. Concerning the option of University level, participants reported more than 60% in all these profession groups and there was no profession effect (see details in Table 8.19).

Table 8.19. Levels of research projects by profession.

Item	Teaching administrators N(%)		Basic studies teaching staff N(%)		Public health teaching staff N(%)		Clinical teaching staff N(%)		χ^2	P
National	15	(16.5)	76	(18.6)	14	(12.5)	63	(18.9)	2.72	0.437
Provincial	51	(56.0)	201	(49.1)	50	(44.6)	187	(56.3)	6.83	0.077
Municipal	15	(16.5)	71	(17.4)	23	(20.5)	125	(37.7)	46.02	<0.001
University	67	(73.6)	256	(62.4)	81	(72.3)	217	(65.2)	6.69	0.083

Influence of Job Title

With regard to the option of National level, all job title groups reported less than 30% and there was no difference among job title groups. Concerning the option of Provincial level, the significant job title effect indicated that Senior group reported highest and Middle group reported lowest. As to the option of Municipal level, there was significant job title effect, with Junior group reporting highest (30.4%) and Middle group reporting lowest (19.3%). Concerning the option of University level, participants reported more than 60% in all these job title groups and there was no difference among them (see details in Table 8.20).

Table 8.20. Levels of research projects by job title.

Item	Junior N(%)		Middle N(%)		Senior N(%)		χ^2	P
National	37	(20.4)	50	(16.9)	81	(17.3)	1.11	0.574
Provincial	98	(54.1)	133	(44.9)	258	(55.2)	8.21	0.016
Municipal	55	(30.4)	57	(19.3)	122	(26.1)	8.35	0.015
University	113	(62.4)	188	(63.1)	320	(68.5)	3.41	0.182

Influence of University's Geographical Location

There were less than 20% reporting on the research projects at National level in both East and West groups and there was no significant difference between these two groups. The difference of involvement in the research projects at Provincial level between East (53.3%) and West (45.8%) groups was not significant. As to the option of Municipal level, East group (28.5%) reported higher than West group (10.0%). Both East and West groups reported more than 60% on University level research projects, and there was no significant difference between them (see Table 8.21).

Table 8.21. Levels of research projects by university's geographical location.

Item	East N(%)		West N(%)		χ^2	*P*
National	138	(18.3)	30	(15.8)	0.64	0.423
Provincial	402	(53.3)	87	(45.8)	3.44	0.064
Municipal	215	(28.5)	19	(10.0)	27.90	<0.001
University	491	(65.0)	130	(68.1)	0.62	0.431

8.4 Publication Number

Regarding the question for staff's publication, they were asked about the number of articles they published on medical teaching and learning research. There were four options participants could choose: None (coded as 1), 1–5 (coded as 2), 6–10 (coded as 3), more than 10 (coded as 4), (see survey Q47 Appendix 1).

Overall findings

The descriptive data to this questions was presented in Table 8.22. The mean of participants' publication was 1.71 which indicating the average number of publication was less than 5 in the whole sample.

Influence of Gender

With regard to their publications on teaching and learning research, male group reported higher than female group (see Table 8.23).

Table 8.22. Publication number.

Item	*N*	*M*	*SD*
How many articles on medical teaching and learning research have you published?	1716	1.71	0.73

Table 8.23. Publication number by gender.

Item	Gender	*N*	*M*	*SD*	*t*	*P*
How many articles on medical teaching and learning research have you published?	Male	759	1.77	0.76	3.20	0.001
	Female	922	1.66	0.69		

Influence of Age

To the publications on medical teaching and learning research, there are significant age effect (see Table 8.24).

Table 8.24. Publication number by age.

Item	Age group	*N*	*M*	*SD*	*F*	*P*
How many articles on medical teaching and learning research have you published?	19–30	394	1.44	0.56	38.33	<0.001
	31–35	410	1.64	0.63		
	36–43	412	1.77	0.78		
	>43	413	1.95	0.77		

The *P*-values of post hoc tests between any two age groups are shown in Table 8.25. The age factor showed a kind of linear effect on the publications on medical teaching and learning research, and the difference between any two age groups was significant. Group 19–30 reported lowest and Group above 43 reported highest. The older the participant's age, the more they published.

Table 8.25. Publication number by age-post hoc analysis.

Item	19–30 vs 31–35	19–30 vs 36–43	19–30 vs >43	31–35 vs 36–43	31–35 vs >43	36–43 vs >43
How many articles on medical teaching and learning research have you published?	<0.001	<0.001	<0.001	0.008	<0.001	<0.001

Influence of Profession

To the publications on medical teaching and learning research, there was significant difference among these four profession groups (see Table 8.26).

The *P*-values of post hoc tests between any two profession groups are shown in Table 8.27. Regarding the publications on medical teaching and learning research, Teaching administrators published more than Basic studies

Table 8.26. Publication number by profession.

Item	Profession	*N*	*M*	*SD*	*F*	*P*
How many articles on medical teaching and learning research have you published?	Teaching administrators	125	1.94	0.77	12.51	<0.001
	Basic studies teaching staff	715	1.77	0.68		
	Public health teaching staff	211	1.73	0.73		
	Clinical teaching staff	665	1.59	0.75		

Table 8.27. Publication number by profession-post hoc analysis.

Item	P1 vs P2	P1 vs P3	P1 vs P4	P2 vs P3	P2 vs P4	P3 vs P4
How many articles on medical teaching and learning research have you published?	0.012	0.010	<0.001	0.538	<0.001	0.010

Note: P1, Teaching administrators; P2, Basic studies teaching staff; P3, Public health teaching staff, P4, Clinical teaching staff.

teaching staff and Public health teaching staff, and these two profession groups published more than Clinical teaching staff.

Influence of Job Title

Concerning the publications on medical teaching and learning research, there was significant difference among these three job title groups (see Table 8.28).

The P-values of post hoc tests between any two title groups are shown in Table 8.29. To the publications on medical teaching and learning research, Senior group reported higher than Middle group, which in turn reported higher than Junior group.

Influence of University's Geographical Location

Regarding the publications on medical teaching and learning research, West group reported higher than East group (see Table 8.30).

Table 8.28. Publication number by job title.

Item	Job title	*N*	*M*	*SD*	*F*	*P*
How many articles on medical teaching and learning research have you published?	Junior	410	1.41	0.57	62.02	<0.001
	Middle	581	1.69	0.65		
	Senior	725	1.89	0.80		

Table 8.29. Publication number by job title-post hoc analysis.

Item	Junior vs Middle	Junior vs Senior	Middle vs Senior
How many articles on medical teaching and learning research have you published?	<0.001	<0.001	<0.001

Table 8.30. Publication number by university's geographical location.

Item	Location	*N*	*M*	*SD*	*t*	*P*
How many articles on medical teaching and learning research have you published?	East	1385	1.69	0.72	−2.60	0.009
	West	331	1.80	0.72		

8.5 Summary

This chapter aims to get general ideas on institutional support for staff's participation in educational research and reform. Four parts are structured in this chapter. The first part reports participants' opinions on the necessity of establishing a unit focused on medicine and health education research. The second part reports if participants' universities had special units or institutions focused on medicine and health education research, and if participants were involved in educational research projects. The third part follows the last question in the second part and intends to get detailed data on the level of research projects they were involved in. The fourth part aims to gain knowledge about staff's publication on medical teaching and learning research. Data analyses in this chapter do not only include overall findings to the above addressed issues, but also take into consideration the impact of gender, age, profession, job title, and the university's geographical locations respectively.

Five questions were asked in correspond with the above mentioned aspects. There were one 5-point Likert-type question, two *'Yes/No'* questions, one multiple-choice question, and one single-choice question. Overall findings to these questions were reported in the four sections in this chapter, and variables such as gender, age, profession, job title, and the university's geographical location were respectively taken into consideration of analysis.

Overall findings in this chapter can be summarized in the following:

1. In general, participants held positive attitude to the necessity of establishing a special research unit or institution focused on medicine and health education. Significant differences were identified in the variables of gender (with male group higher than female group), age (with youngest and eldest groups higher than the middle), profession (with Teaching administrators higher than other groups of teaching staff), and university's geographical locations (with East group higher than West group).
2. The majority of the participants reported that their universities already had special research units or institutions focusing on medicine and health education research. Half of these participants had been or currently were involved in educational research projects. Concerning the involvement in educational research,

significant differences were identified in terms of gender (with male group higher than female group), age (with eldest group higher than other groups), job title (with Senior group higher than Middle and Junior groups), profession (with Teaching administrators highest and Clinical teaching staff lowest), and university's geographical locations (with East group higher than West group).

3. Less than 30% participants reported their involvement in research projects at National and Municipal levels. More than 50% participants reported their involvement in Provincial and University level research projects.
4. The mean of participants' publication was 1.71 meaning the average number of publication was less than 5 in the whole sample. Significant differences were identified in terms of gender (with male group higher than female group), age (elder groups higher than younger groups), job title (Senior group higher than Middle and Junior groups), profession (with Teaching administrators highest and Clinical teaching staff lowest), and university's geographical locations (with West group higher than East group).

To conclude, a strong agreement on the necessity in establishing special research units or institutions focusing on medicine and health education research was reached by the participants in the present survey. The findings show that more than half of the investigated universities have already established such a research unit, although they are mostly located in the East group. Only one university from the West group has such an organization. About half of the participants reported their involvement in educational research projects and most projects were at the levels of Provincial and University. These results also suggests an uneven distribution of the groups in relation to their participation in educational research projects as well as the amount of relevant publications, for example, male group more than female group, elder groups more than younger groups, Senior groups more than Junior groups, Teaching administrators more than Basic studies teaching staff and Public health teaching staff, while Clinical teaching staff lowest. One interesting result is despite their less resources for research projects, participants from West group reported higher amount of publications than those from East group. All in all, this indicates that although there is an overall positive situation with regards to institutional support for educational research in medicine and health, the current involvement is limited to certain groups of teaching staff. For further reform, there is a need to provide sufficient support at an institutional level to all groups of teaching staff.

9

Staff's Opinion on General Development of Medicine and Health Education

Xiaoju Duan, Jiannong Shi and Xiangyun Du

Abstract

This chapter aims to gain an overall knowledge about staff's attitudes towards general development of medicine and health education. It also intends to have an overview about the participants' familiarity with the Global Standards of Undergraduate Medical Education, and the accreditation of clinical medicine program conducted by the Ministry of Education of China. Eleven questions were asked with regard to these aspects in the survey. Findings are presented in this chapter, which are structured in two sections. Section 1 reports how respondents scaled the importance of including professional morality, attitude, behavior, and ethics in medicine and health education; to what extent they think that it is necessary to provide specific course and training activities on how to develop medical communication skills; to what extent they think it is important for medicine graduates-to-be to have knowledge of ethics and legislation issues in the profession; to what extent they think medical informatics courses should be provided in clinical medicine program; to what extent they think evidence-based medicine courses should be provided in clinical medicine program; to

Change and Reform in Medicine and Health Education in China, 157–174.

what extent they think that the current medicine curricula in China should be reformed; to what extent they believe that the current medicine curricula contents should be updated; to what extent they think that the current continued education system for medical practitioners should be improved; to what extent they are satisfied with the current medicine and health education system in China. Section 2 reports participants' familiarity with the Global Standards of Undergraduate Medical Education, and the accreditation of clinical medicine program conducted by the Ministry of Education. Data analyses in this chapter do not only include overall findings to the above addressed issues, but also take into consideration of the impact of gender, age, profession, job title, and the university's geographical location respectively.

9.1 Perception on the General Development of Medicine and Health Education

In order to understand teaching staff's overall perception on the general development of medicine and health education, nine questions were asked that require 5-point Likert-type answers. These questions include: (1) To what extent they think that it is important to include professional morality, attitude, behavior, and ethics in medicine and health education (see survey Q48 Appendix 1), (2) To what extent they think it is necessary to provide specific course and training activities on how to develop medical communication skills (see survey Q49 Appendix 1), (3) To what extent they think it is important for medicine graduates-to-be to have knowledge of ethics and legislation issues in the profession (see survey Q50 Appendix 1), (4) To what extent they think medical informatics courses should be provided in clinical medicine program (see survey Q51 Appendix 1), (5) To what extent they think evidence-based medicine courses should be provided in clinical medicine program (see survey Q52 Appendix 1), (6) To what extent they think that the current medicine curricula in China should be reformed (see survey Q55 Appendix 1), (7) To what extent they believe that the current medicine curricula contents should be updated, (see survey Q56 Appendix 1), (8) To what extent they think that the current continued education system for medical practitioners should be improved (see survey Q57 Appendix 1), and (9) To what extent they are satisfied with the current medicine and health education system in China (see survey Q58 Appendix 1).

Overall findings

In general, participants had positive attitude to all these nine questions, especially the first one, the importance of including professional morality, attitude, behavior, and ethics in medicine and health education (see Table 9.1).

Table 9.1. Perception on the general development of medicine and health education.

Item	*N*	*M*	*SD*
To what extent do you think it is important to include professional morality, attitude, behavior, and ethics in medicine and health education?	1795	4.58	0.56
To what extent do you think it is necessary to provide specific course and training activities on how to develop medical communication skills?	1781	4.30	0.70
To what extent do you think it is important for medicine graduates-to-be to have knowledge of ethics and legislation issues in the profession?	1791	4.46	0.60
To what extent do you think medical informatics courses should be provided in in clinical medicine program?	1793	4.32	0.66
To what extent do you think evidence-based medicine courses should be provided in clinical medicine program?	1785	4.31	0.69
To what extent do you think that the current medicine curricula in China should be reformed?	1785	4.31	0.65
To what extent do you believe that the current medicine curricula contents should be updated?	1791	4.28	0.67
To what extent do you think that the current continued education system for medical practitioners should be improved?	1791	4.38	0.66
To what extent are you satisfied with the current medicine and health education system in China?	1792	3.48	0.93

Influence of Gender

Both male and female groups had positive attitude to all these nine questions, namely, the importance of including professional morality, attitude, behavior, and ethics in medicine and health education, the necessity of providing specific course and training activities on how to develop medical communication skills, the importance for medicine graduates-to-be to have knowledge of ethics and legislation issues in the profession, the inclusion of medical informatics courses in clinical medicine program, the inclusion of evidence-based medicine courses in clinical medicine program, the reform of current medicine curricula in China, the update of current medicine curricula, the improvement of current continued education system for medical practitioners, and the satisfaction with the current medicine and health education system in China. Male and female groups had similar opinions on these nine questions (see Table 9.2).

Table 9.2. Perception on the general development of medicine and health education by gender.

Item	Gender	*N*	*M*	*SD*	*t*	*P*
To what extent do you think it is important to include professional morality, attitude, behavior, and ethics in medicine and health education?	Male	801	4.57	0.58	−0.54	0.591
	Female	954	4.58	0.55		
To what extent do you think it is necessary to provide specific course and training activities on how to develop medical communication skills?	Male	792	4.31	0.69	0.74	0.459
	Female	950	4.29	0.70		
To what extent do you think it is important for medicine graduates-to-be to have knowledge of ethics and legislation issues in the profession?	Male	796	4.46	0.59	−0.05	0.957
	Female	955	4.46	0.61		
To what extent do you think medical informatics courses should be provided in clinical medicine program?	Male	799	4.32	0.67	0.05	0.961
	Female	954	4.32	0.66		
To what extent do you think evidence-based medicine courses should be provided in clinical medicine program?	Male	794	4.32	0.69	0.47	0.639
	Female	952	4.30	0.70		
To what extent do you think that the current medicine curricula in China should be reformed?	Male	799	4.32	0.66	1.17	0.242
	Female	950	4.29	0.64		
To what extent do you believe that the current medicine curricula contents should be updated?	Male	798	4.29	0.68	0.84	0.403
	Female	955	4.26	0.66		
To what extent do you think that the current continued education system for medical practitioners should be improved?	Male	800	4.37	0.68	−0.52	0.601
	Female	953	4.38	0.64		
To what extent are you satisfied with the current medicine and health education system in China?	Male	798	3.50	0.98	0.59	0.556
	Female	955	3.47	0.89		

Influence of Age

Participants from different age groups had positive attitude to all these nine questions. Participants from these four age groups held similar opinions on the importance of including professional morality, attitude, behavior, and ethics in medicine and health education, the importance for medicine graduates-to-be to have knowledge of ethics and legislation issues in the profession, the inclusion of medical informatics courses in clinical medicine program, the inclusion of evidence-based medicine courses in clinical medicine program, the reform of current medicine curricula in China, the update of current medicine curricula, the satisfaction with the current medicine and health education system in China. There were significant differences among the four age groups on the necessity of providing specific course and training activities on how to develop medical communication skills, the improvement of current continued education system for medical practitioners (see Table 9.3).

Table 9.3. Perception on the general development of medicine and health education by age.

Item	Age group	*N*	*M*	*SD*	*F*	*P*
To what extent do you think it is important to include professional morality, attitude, behavior, and ethics in medicine and health education?	19–30	408	4.58	0.56	1.08	0.355
	31–35	426	4.54	0.60		
	36–43	428	4.58	0.56		
	> 43	432	4.61	0.54		
To what extent do you think it is necessary to provide specific course and training activities on how to develop medical communication skills?	19–30	405	4.39	0.65	3.30	0.020
	31–35	424	4.27	0.73		
	36–43	426	4.31	0.70		
	> 43	426	4.25	0.71		
To what extent do you think it is important for medicine graduates-to-be to have knowledge of ethics and legislation issues in the profession?	19–30	408	4.50	0.58	0.67	0.570
	31–35	424	4.44	0.64		
	36–43	426	4.46	0.61		
	> 43	432	4.46	0.61		
To what extent do you think medical informatics courses should be provided in clinical medicine program?	19–30	407	4.39	0.64	2.51	0.057
	31–35	425	4.26	0.67		
	36–43	428	4.31	0.68		
	> 43	432	4.31	0.67		
To what extent do you think evidence-based medicine courses should be provided in clinical medicine program?	19–30	407	4.35	0.68	0.70	0.549
	31–35	422	4.28	0.70		
	36–43	427	4.31	0.71		
	> 43	429	4.30	0.71		
To what extent do you think that the current medicine curricula in China should be reformed?	19–30	408	4.28	0.67	0.37	0.773
	31–35	426	4.30	0.64		
	36–43	428	4.33	0.67		
	> 43	426	4.31	0.64		
To what extent do you believe that the current medicine curricula contents should be updated?	19–30	408	4.29	0.64	1.59	0.191
	31–35	427	4.22	0.68		
	36–43	427	4.31	0.69		
	> 43	430	4.30	0.67		
To what extent do you think that the current continued education system for medical practitioners should be improved?	19–30	408	4.35	0.67	5.09	0.002
	31–35	425	4.29	0.68		
	36–43	428	4.39	0.67		
	> 43	431	4.46	0.63		
To what extent are you satisfied with the current medicine and health education system in China?	19–30	408	3.54	0.87	1.23	0.298
	31–35	426	3.46	0.93		
	36–43	428	3.42	0.99		
	> 43	430	3.50	0.94		

The *P*-value of post hoc tests between any two age groups are shown in Table 9.4. With regard to the necessity of providing specific course and training activities on how to develop medical communication skills, Group 19–30 reported significantly higher than Group 31–35 and Group 36–43. As to the improvement of current continued education system for medical practitioners, Group above 43 reported higher than Group 19–30 and Group 31–35, and Group 36–43 reported higher than Group 31–35.

Table 9.4. Perception on the general development of medicine and health education by age-post hoc analysis.

Item	19–30 vs 31–35	19–30 vs 36–43	19–30 vs >43	31–35 vs 36–43	31–35 vs >43	36–43 vs >43
To what extent do you think it is important to include professional morality, attitude, behavior, and ethics in medicine and health education?	0.269	0.923	0.512	0.307	0.075	0.446
To what extent do you think it is necessary to provide specific course and training activities on how to develop medical communication skills?	0.012	0.079	0.004	0.449	0.675	0.239
To what extent do you think it is important for medicine graduates-to-be to have knowledge of ethics and legislation issues in the profession?	0.179	0.343	0.294	0.689	0.761	0.922
To what extent do you think medical informatics courses should be provided in clinical medicine program?	0.007	0.075	0.110	0.351	0.259	0.847
To what extent do you think evidence-based medicine courses should be provided in clinical medicine program?	0.155	0.501	0.345	0.447	0.625	0.784
To what extent do you think that the current medicine curricula in China should be reformed?	0.800	0.319	0.573	0.452	0.753	0.662
To what extent do you believe that the current medicine curricula contents should be updated?	0.152	0.595	0.775	0.047	0.082	0.802
To what extent do you think that the current continued education system for medical practitioners should be improved?	0.165	0.415	0.017	0.026	<0.001	0.112
To what extent are you satisfied with the current medicine and health education system in China?	0.208	0.072	0.544	0.585	0.507	0.226

Influence of Profession

Participants from all these four profession groups had positive attitude to all these nine questions. There were significant differences among profession groups on eight out of these nine items, namely, the importance of including professional morality, attitude, behavior, and ethics in medicine and health education, the necessity of providing specific course and training activities on how to develop medical communication skills, the importance for medicine graduates-to-be to have knowledge of ethics and legislation issues in the profession, the inclusion of medical informatics courses in clinical medicine program, the inclusion of evidence-based medicine courses in clinical medicine program, the reform of current medicine curricula in China, the update of current medicine curricula, the improvement of current continued education system for medical practitioners. Participants of different profession had similar opinions on the satisfaction with the current medicine and health education system in China (see Table 9.5).

The P-values of post hoc tests between any two profession groups are shown in Table 9.6. For the importance of including professional morality, attitude, behavior, and ethics in medicine and health education, Teaching administrators and Clinical teaching staff reported higher than Public health teaching staff; Clinical teaching staff reported higher than Basic studies teaching staff. As to the necessity of providing specific course and training activities on how to develop medical communication skills, Teaching administrators and Clinical teaching staff reported higher than Basic studies teaching staff and Public health teaching staff. With regard to the importance for medicine graduates-to-be to have knowledge of ethics and legislation issues in the profession, Teaching administrators and Clinical teaching staff reported higher than Basic studies teaching staff and Public health teaching staff. For the inclusion of medical informatics courses in clinical medicine program, Teaching administrators and Clinical teaching staff reported higher than Basic studies teaching staff and Public health teaching staff. To the inclusion of evidence-based medicine courses in clinical medicine program, Teaching administrators, Public health teaching staff, and Clinical teaching staff reported higher than Basic studies teaching staff. Regarding the reform of current medicine curricula in China, Clinical teaching staff reported higher than Basic studies teaching staff. As to the update of current medicine curricula, Teaching administrators and Clinical teaching staff reported higher than Basic studies teaching staff. For the improvement of current continued education system for medical practitioners, Teaching administrators and Clinical teaching staff reported higher than Basic studies teaching staff.

Table 9.5. Perception on the general development of medicine and health education by profession.

Item	Profession	*N*	*M*	*SD*	*F*	*P*
To what extent do you think it is important to include professional morality, attitude, behavior, and ethics in medicine and health education?	Teaching administrators	127	4.64	0.56	6.02	<0.001
	Basic studies teaching staff	736	4.55	0.59		
	Public health teaching staff	219	4.47	0.58		
	Clinical teaching staff	713	4.63	0.53		
To what extent do you think it is necessary to provide specific course and training activities on how to develop medical communication skills?	Teaching administrators	127	4.48	0.60	16.83	<0.001
	Basic studies teaching staff	725	4.22	0.72		
	Public health teaching staff	219	4.13	0.77		
	Clinical teaching staff	710	4.41	0.64		
To what extent do you think it is important for medicine graduates-to-be to have knowledge of ethics and legislation issues in the profession?	Teaching administrators	127	4.53	0.57	10.73	<0.001
	Basic studies teaching staff	729	4.39	0.63		
	Public health teaching staff	219	4.37	0.69		
	Clinical teaching staff	716	4.55	0.54		
To what extent do you think medical informatics courses should be provided in clinical medicine program?	Teaching administrators	126	4.40	0.63	11.14	<0.001
	Basic studies teaching staff	732	4.24	0.69		
	Public health teaching staff	219	4.21	0.74		
	Clinical teaching staff	716	4.42	0.59		
To what extent do you think evidence-based medicine courses should be provided in clinical medicine program?	Teaching administrators	127	4.46	0.57	20.92	<0.001
	Basic studies teaching staff	727	4.16	0.72		
	Public health teaching staff	216	4.33	0.74		
	Clinical teaching staff	715	4.43	0.64		
To what extent do you think that the current medicine curricula in China should be reformed?	Teaching administrators	126	4.36	0.70	4.69	0.003
	Basic studies teaching staff	729	4.24	0.66		
	Public health teaching staff	218	4.29	0.60		
	Clinical teaching staff	712	4.37	0.64		
To what extent do you believe that the current medicine curricula contents should be updated?	Teaching administrators	127	4.39	0.62	7.16	<0.001
	Basic studies teaching staff	731	4.20	0.67		
	Public health teaching staff	218	4.28	0.64		
	Clinical teaching staff	715	4.34	0.67		
To what extent do you think that the current continued education system for medical practitioners should be improved?	Teaching administrators	127	4.46	0.60	3.68	0.012
	Basic studies teaching staff	730	4.32	0.64		
	Public health teaching staff	218	4.36	0.67		
	Clinical teaching staff	716	4.42	0.68		
To what extent are you satisfied with the current medicine and health education system in China?	Teaching administrators	127	3.50	0.83	0.18	0.911
	Basic studies teaching staff	730	3.49	0.88		
	Public health teaching staff	219	3.44	0.87		
	Clinical teaching staff	716	3.49	1.02		

Table 9.6. Perception on the general development of medicine and health education by profession-post hoc analysis.

Item	P1 vs P2	P1 vs P3	P1 vs P4	P2 vs P3	P2 vs P4	P3 vs P4
To what extent do you think it is important to include professional morality, attitude, behavior, and ethics in medicine and health education?	0.095	0.006	0.861	0.059	0.006	<0.001
To what extent do you think it is necessary to provide specific course and training activities on how to develop medical communication skills?	<0.001	<0.001	0.278	0.089	<0.001	<0.001
To what extent do you think it is important for medicine graduates-to-be to have knowledge of ethics and legislation issues in the profession?	0.019	0.018	0.693	0.626	<0.001	<0.001
To what extent do you think medical informatics courses should be provided in clinical medicine program?	0.016	0.011	0.743	0.495	<0.001	<0.001
To what extent do you think evidence-based medicine courses should be provided in clinical medicine program?	<0.001	0.106	0.662	0.001	<0.001	0.074
To what extent do you think that the current medicine curricula in China should be reformed?	0.067	0.380	0.880	0.309	<0.001	0.145
To what extent do you believe that the current medicine curricula contents should be updated?	0.002	0.125	0.425	0.106	<0.001	0.222
To what extent do you think that the current continued education system for medical practitioners should be improved?	0.025	0.163	0.527	0.440	0.003	0.220
To what extent are you satisfied with the current medicine and health education system in China?	0.880	0.558	0.866	0.509	0.974	0.524

Note: P1, Teaching administrators; P2, Basic studies teaching staff; P3, Public health teaching staff; P4, Clinical teaching staff.

Influence of Job Title

Participants of different job title were positive to all these nine questions. The job title effect was significant on six out of nine questions, namely, the importance of including professional morality, attitude, behavior, and ethics in medicine and health education, the inclusion of medical informatics courses in clinical medicine program, the inclusion of evidence-based medicine courses in clinical medicine program, the reform of current medicine curricula in China, the update of current medicine curricula, the improvement of current continued education system for medical practitioners. Participants of different job title had similar opinions on the necessity of providing specific course and training activities on how to develop medical communication skills, the importance for medicine graduates-to-be to have knowledge of ethics and legislation issues in the profession, the satisfaction with the current medicine and health education system in China (see Table 9.7).

The P-value of post hoc tests between any two title groups are shown in Table 9.8. Regarding the importance of including professional morality, attitude, behavior, and ethics in medicine and health education, Senior group reported higher than Middle group. For the inclusion of medical informatics courses in clinical medicine program, Junior group reported higher than Middle group. With the inclusion of evidence-based medicine courses in clinical medicine program, Junior and Senior groups reported higher than Middle group. As to the reform of current medicine curricula in China, Senior group reported higher than Middle group. With regard to the update of current medicine curricula, Senior group reported higher than Middle group. For the improvement of current continued education system for medical practitioners, Senior group reported higher than Junior and Middle groups.

Influence of University's Geographical Location

Participants displayed positive attitude to all these nine questions in both East and West groups. There were no significant differences between East and West groups on six out of nine questions, namely, the importance of including professional morality, attitude, behavior, and ethics in medicine and health education, the necessity of providing specific course and training activities on

Table 9.7. Perception on the general development of medicine and health education by job title.

Item	Job title	*N*	*M*	*SD*	*F*	*P*
To what extent do you think it is important to include professional morality, attitude, behavior, and ethics in medicine and health education?	Junior	433	4.56	0.57	3.67	0.026
	Middle	607	4.54	0.59		
	Senior	755	4.62	0.54		
To what extent do you think it is necessary to provide specific course and training activities	Junior	433	4.36	0.65	2.54	0.079
	Middle	602	4.27	0.72		
on how to develop medical communication skills?	Senior	746	4.29	0.70		
To what extent do you think it is important for medicine graduates-to-be to have knowledge of ethics and legislation issues in the profession?	Junior	433	4.48	0.58	0.73	0.482
	Middle	604	4.44	0.61		
	Senior	754	4.47	0.61		
To what extent do you think medical informatics courses should be provided in clinical medicine program?	Junior	433	4.39	0.63	4.80	0.008
	Middle	606	4.26	0.68		
	Senior	754	4.33	0.67		
To what extent do you think evidence-based medicine courses should be provided in clinical medicine program?	Junior	432	4.34	0.68	3.47	0.031
	Middle	603	4.25	0.70		
	Senior	750	4.34	0.69		
To what extent do you think that the current medicine curricula in China should be reformed?	Junior	430	4.28	0.66	3.17	0.042
	Middle	607	4.27	0.65		
	Senior	748	4.35	0.64		
To what extent do you believe that the current medicine curricula contents should be updated?	Junior	432	4.27	0.65	5.65	0.004
	Middle	608	4.22	0.68		
	Senior	751	4.34	0.66		
To what extent do you think that the current continued education system for medical practitioners should be improved?	Junior	432	4.35	0.66	8.91	<0.001
	Middle	607	4.30	0.68		
	Senior	752	4.45	0.63		
To what extent are you satisfied with the current medicine and health education system in China?	Junior	433	3.56	0.91	2.28	0.103
	Middle	607	3.44	0.90		
	Senior	752	3.48	0.97		

how to develop medical communication skills, the importance for medicine graduates-to-be to have knowledge of ethics and legislation issues in the profession, the inclusion of evidence-based medicine courses in clinical medicine program, the reform of current medicine curricula in China, the improvement of current continued education system for medical practitioners. East group reported significantly higher than West group on the inclusion of medical informatics courses in clinical medicine program, the update of current medicine curricula, the satisfaction with the current medicine and health education system in China (see Table 9.9).

Table 9.8. Perception on the general development of medicine and health education by job title-post hoc analysis.

Item	Junior vs Middle	Junior vs Senior	Middle vs Senior
To what extent do you think it is important to include professional morality, attitude, behavior, and ethics in medicine and health education?	0.538	0.086	0.009
To what extent do you think it is necessary to provide specific course and training activities on how to develop medical communication skills?	0.029	0.078	0.568
To what extent do you think it is important for medicine graduates-to-be to have knowledge of ethics and legislation issues in the profession?	0.273	0.793	0.330
To what extent do you think medical informatics courses should be provided in clinical medicine program?	0.002	0.108	0.076
To what extent do you think evidence-based medicine courses should be provided in clinical medicine program?	0.032	0.925	0.018
To what extent do you think that the current medicine curricula in China should be reformed?	0.753	0.074	0.019
To what extent do you believe that the current medicine curricula contents should be updated?	0.206	0.089	0.001
To what extent do you think that the current continued education system for medical practitioners should be improved?	0.231	0.013	<0.001
To what extent are you satisfied with the current medicine and health education system in China?	0.034	0.149	0.393

9.2 The Familiarity with New Standards

In order to understand staff's familiarity with new standards and accreditation in medicine and health education, two '*Yes/No*' questions were asked. These questions include: (1) Are they familiar with the Global Standards of Undergraduate Medical Education (see survey Q53 Appendix 1); (2) Are they familiar with the accreditation of clinical medicine program conducted by the Ministry of Education (see survey Q54 Appendix 1).

Overall findings

In general, less than 50% participants reported that they were familiar with Global Standards of Undergraduate Medical Education, and the accreditation

Table 9.9. Perception on the general development of medicine and health education by university's geographical location.

Item	Location	*N*	*M*	*SD*	*t*	*P*
To what extent do you think it is important to include professional morality, attitude, behavior, and ethics in medicine and health education?	East West	1437 358	4.58 4.56	0.56 0.59	0.65	0.515
To what extent do you think it is necessary to provide specific course and training activities on how to develop medical communication skills?	East West	1426 355	4.32 4.25	0.69 0.74	1.71	0.088
To what extent do you think it is important for medicine graduates-to-be to have knowledge of ethics and legislation issues in the profession?	East West	1433 358	4.47 4.41	0.59 0.65	1.72	0.086
To what extent do you think medical informatics courses should be provided in clinical medicine program?	East West	1436 357	4.34 4.25	0.66 0.68	2.17	0.030
To what extent do you think evidence-based medicine courses should be provided in clinical medicine program?	East West	1427 358	4.32 4.25	0.69 0.72	1.85	0.064
To what extent do you think that the current medicine curricula in China should be reformed?	East West	1429 356	4.31 4.28	0.63 0.71	1.01	0.311
To what extent do you believe that the current medicine curricula contents should be updated?	East West	1435 356	4.30 4.21	0.65 0.75	2.16	0.031
To what extent do you think that the current continued education system for medical practitioners should be improved?	East West	1433 358	4.39 4.35	0.64 0.71	0.86	0.392
To what extent are you satisfied with the current medicine and health education system in China?	East West	1436 356	3.52 3.34	0.93 0.92	3.22	0.001

of clinical medicine program conducted by the Ministry of Education (see Table 9.10).

Influence of Gender

Male group (50.0%) reported significantly higher percentage than female group (39.9%) did on the familiarity with the Global Standards of Undergraduate Medical Education. With regard to the familiarity with the accreditation of clinical medicine program conducted by the Ministry of Education, male group (53.5%) also reported higher than female group (45.5%) (see Table 9.11).

Table 9.10. The familiarity with new standards.

Item	*N*	Yes N/(%)	No N/(%)
Are you familiar with the Global Standards of Undergraduate Medical Education?	1789	794 (44.4)	995 (55.6)
Are you familiar with the accreditation of clinical medicine program conducted by the Ministry of Education?	1771	867 (49.0)	904 (51.0)

Table 9.11. The familiarity with new standards by gender.

Item	Male N/(%)		Female N/(%)		χ^2	*P*
Are you familiar with the Global Standards of Undergraduate Medical Education?	399	(50.0)	380	(39.9)	17.87	<0.001
Are you familiar with the accreditation of clinical medicine program conducted by the Ministry of Education?	424	(53.5)	428	(45.5)	11.16	0.001

Influence of Age

As to the familiarity with the Global Standards of Undergraduate Medical Education, the age effect was significant, with Group above 43 reporting highest and Group 19–30 reporting lowest. With regard to familiarity with the accreditation of clinical medicine program conducted by the Ministry of Education, the significant age effect indicated that Group above 43 reported highest and Group 19–30 reported lowest (see Table 9.12).

Table 9.12. The familiarity with new standards by age.

Item	19–30 N/(%)		31–35 N/(%)		36–43 N/(%)		>43 N/(%)		χ^2	*P*
Are you familiar with the Global Standards of Undergraduate Medical Education?	156	(38.3)	164	(38.5)	203	(47.5)	221	(51.5)	22.55	<0.001
Are you familiar with the accreditation of clinical medicine program conducted by the Ministry of Education?	169	(41.8)	203	(48.3)	209	(49.4)	233	(54.6)	13.62	0.003

Influence of Profession

Concerning the familiarity with the Global Standards of Undergraduate Medical Education, there was significant difference among these four profession groups, with Teaching administrators reporting highest (70.9%) and Basic studies teaching staff reporting lowest (40.1%). With the familiarity with the accreditation of clinical medicine program conducted by the Ministry of Education, the profession effect was significant, with Teaching administrators reporting highest and Public health teaching staff reporting lowest (see details in Table 9.13).

Table 9.13. The familiarity with new standards by profession.

Item	Teaching administrators N(%)		Basic studies teaching staff N(%)		Public health teaching staff N(%)		Clinical teaching staff N(%)		χ^2	*P*
Are you familiar with the Global Standards of Undergraduate Medical Education?	90	(70.9)	294	(40.1)	80	(36.9)	330	(46.3)	47.59	<0.001
Are you familiar with the accreditation of clinical medicine program conducted by the Ministry of Education?	83	(65.9)	333	(46.2)	75	(34.4)	376	(53.3)	40.35	<0.001

Influence of Job Title

For the familiarity with the Global Standards of Undergraduate Medical Education, there was significant job title effect, with Senior group reporting highest and Middle group reporting lowest. With the familiarity with the accreditation of clinical medicine program conducted by the Ministry of Education, the difference among these three job title groups was significant, with Senior group reporting highest and Junior group reporting lowest (see details in Table 9.14).

Influence of University's Geographical Location

Participants from both East and West groups reported less than 50% on the familiarity with the Global Standards of Undergraduate Medical Education,

Table 9.14. The familiarity with new standards by job title.

Item	Junior N/(%)		Middle N/(%)		Senior N/(%)		χ^2	*P*
Are you familiar with the Global Standards of Undergraduate Medical Education?	175	(40.6)	238	(39.1)	381	(50.8)	21.76	< 0.001
Are you familiar with the accreditation of clinical medicine program conducted by the Ministry of Education?	188	(43.8)	269	(45.0)	410	(55.1)	19.57	< 0.001

and the familiarity with the accreditation of clinical medicine program conducted by the Ministry of Education. There was no significant differences between East and West groups on these two questions (see Table 9.15).

Table 9.15. The familiarity with new standards by university's geographical location.

Item	East N/(%)		West N/(%)		χ^2	*P*
Are you familiar with the Global Standards of Undergraduate Medical Education?	640	(44.7)	154	(43.1)	0.28	0.597
Are you familiar with the accreditation of clinical medicine program conducted by the Ministry of Education?	705	(49.8)	162	(45.5)	2.12	0.145

9.3 Summary

The purpose of this chapter is to understand staff's overall opinions towards general development of medicine and health development. This chapter also intends to get general ideas about participants' familiarity with the Global Standards of Undergraduate Medical Education, and their familiarity with the accreditation of clinical medicine program conducted by the Ministry of Education of China. Accordingly, there are two sections in the present chapter. First, it shows participants' opinions on the importance of including professional morality, attitude, behavior, and ethics in medicine and health education, the necessity of providing specific course and training activities on how to develop medical communication skills, the importance for medicine graduates-to-be to have knowledge of ethics and legislation issues in the profession, the inclusion of medical informatics courses in clinical medicine program, the inclusion of evidence-based medicine courses in clinical medicine program, the reform of current medicine curricula in China, the update of current medicine curricula, the improvement of current continued education system for medical practitioners, the satisfaction with the current medicine and health education system in China. Second, staff's response on whether they are familiar with new educational standards are provided.

Eleven questions were asked with regards to these two aspects in the survey, which included nine 5-point Likert-type questions and two '*Yes/No*' questions. Overall findings to these questions were reported in the two sections in this chapter, and variables such as gender, age, profession, job title, and the university's geographical location were respectively taken into consideration of analysis.

Overall findings in this chapter can be summarized in the following:

1. Participants in this investigation generally agreed on the importance of including professional morality, attitude, behavior, and ethics in medicine and health education, the necessity of providing specific course and training activities on how to develop medical communication skills, the importance for medicine graduates-to-be to have knowledge of ethics and legislation issues in the profession, the inclusion of medical informatics courses in clinical medicine program, the inclusion of evidence-based medicine courses in clinical medicine program, the reform of current medicine curricula in China, the update of current medicine curricula, the improvement of current continued education system for medical practitioners, the satisfaction with the current medicine and health education system in China. Significant differences were identified in these aspects in terms of profession (with Teaching administrators and Clinical teaching staff generally higher than Basic studies teaching staff and Public health teaching staff) and job titles (with Middle group lower than Senior and Junior groups).
2. With regard to the Global Standards of Undergraduate Medical Education and the accreditation of clinical medicine program conducted by the Ministry of Education, less than half participants reported their familiarity. Significant differences were identified in terms of gender (with male group higher than female group), age (with elder groups higher than younger groups), profession (with Teaching administrators and Clinical teaching staff generally

higher than Basic studies teaching staff and Pubic health teaching staff), and job titles (with Senior group higher than Middle and Junior groups).

In conclusion, teaching staff in this study generally held positive attitude to the general development of medicine and health education; however they have limited familiarity with the new trend of change within medicine and health education both worldwide and nationwide. The general knowledge about these trends and standards are mainly restricted to certain groups of teaching staff, which indicates the limitation of the institutional support to facilitate further educational reform.

10

Towards Further Change in Medicine and Health Education in China

Xiangyun Du, Jiannong Shi and Baozhi Sun

10.1 Discussion of Research Results

10.1.1 Summary of the Study

In earlier chapters of this book, we introduced the background of this research project: the global and national background of educational reforms and change in medicine and health education, why we focus on teaching staff's perspective, and how the investigation was conducted.

The body of this book discussed the research findings of this study. A survey-based investigation (see Chapter 3 for details of the survey) was conducted during late 2009 among 1955 members of teaching staff from 22 medical universities in China, netting 1820 (93.1%) respondents. This survey was carried out in the form of a questionnaire that consisted of six themes with 58 questions. These themes included staff's attitudes toward and involvement in the following aspects: (1) general reforms in medicine and health education such as curricula integration, (2) pedagogy innovation such as employment of Problem-Based Learning methods (PBL), (3) new assessment methods, (4) educational technologies in the teaching practice, (5) institutional support

Change and Reform in Medicine and Health Education in China, 175–188.

for teaching staff's participation in educational research and reform, (6) further development of medicine and health education. Results of the study are reported in Chapters 4–9 of the book.

In this chapter, we will discuss the results in relation to the research questions posed in Chapter 1, as well as related context and theories as discussed in Chapter 2 with regard to how educational change should take place and the important role teaching staff should play in the change's implementation. We will conclude this chapter and book with recommendations on further steps toward reform as well as future research work in this field.

10.1.2 Teaching Staff's Attitudes Towards Reforms

In general, teaching staff who have participated in this investigation, despite differences in gender, age, profession, job title, and geographical location of their university, agreed on the significant importance of educational reforms in medicine and health education in China. They reported positive attitudes towards the reforms conducted, including reforms in such varied areas as curricula integration, adoption of innovative teaching methods such as PBL, employment of new assessment methods for students' learning and for evaluating teaching outcomes, use of educational technologies in teaching practices, engagement in educational research that is related to their teaching experiences, and support for strengthening new skills and competences (such as communication skills, ethical thinking, and professionalism, among others) to be included in the required educational outcomes. The overall positive feelings towards reforms in these areas transcended differentiating factors such as gender, age, profession, job title, and geographical location of their universities. These findings show a positive overall picture of staff's support for educational reforms in medicine and health in China.

The results of this study also evidenced that there is a need for further reforms and change in medicine and health education in China from teaching staff's perspective. Nevertheless, as reported in Chapters 4–9, factors such as age, profession, job title, and geographical location of their universities made a difference in the participants' attitudes towards concrete aspects of the reforms.

Studies on educational change very often report their observation of some staff/faculty who are against reforms (Bloom, 1989; Cantor *et al.*, 1991;

Mennin and Krackov, 1998; Alarcao, 2007; Frenay *et al.*, 2007; Stojcevski *et al.*, 2012). In this study, although it was, in part, born because of concerns from resistance to change (as introduced in Chapter 1) (Sun, 2005; 2011), the results reported a general support for reforms from teaching staff. While we acknowledge that resistance still exists, the positive results in this study indicate that it is likely reform measures will be passed.

10.1.3 Teaching Staff's Involvement and Participation in Reforms

In addition to their positive attitudes, teaching staff participating in this study also reported a certain level of knowledge of specific reforms, such as curricula integration, new teaching and assessment methods, and use of certain educational technologies. Nevertheless, bearing in mind that the pilot stage of reforms had already been conducted in these universities, their reported knowledge was at a rather limited level. In particular, although some of the teaching staff claimed their participation in innovation in teaching and assessment methods, a general lack of deep understanding of student-centered learning could be identified. This limitation in understanding would be a barrier in developing skills for conducting reforms so as to maximize students' learning outcomes (Ludmerer, 2004). In addition, results show that participation in educational research is only limited to certain groups of teaching staff in this study. Less than half of the participants reported their familiarity with different global standards of medicine and health education and the accreditation of the clinical medicine program conducted by the Ministry of Education (MOE, 2005). Excepting groups of Teaching administrators and Clinical teaching staff, interpersonal communication skills, is not, thus far, an area acknowledged by the teaching staff in this study. This skill is an important learning outcome suggested in the ACGME Core Competencies (ACGME, 1999), Global Minimum Essential Requirements (IIME Core Committee, 2002), Tomorrow's Doctors (GMC, 2009), and the Tuning project (Cumming and Ross, 2008), as well as others in the Western world. It is also emphasized in the accreditation requirement of the clinical medicine program in China (MOE, 2005).

This investigation also took into consideration factors such as gender, age, profession, job title, and geographical location of the respondents' universities. The results evidenced significant differences among groups in relation to their knowledge, skills, and access, in relation to the aspects being researched.

This study identified that more participants with Junior and Senior job titles support educational reform. In relation to this, a 'U' shaped tendency of attitudes towards educational reform in medicine and health education was recognized, in that Junior and Senior teaching staff are more supportive for educational reform than teaching staff with Middle job titles.

Teaching staff's professions also made a significant impact on their attitudes towards reforms and their involvement therein. Particularly, Teaching administrators and Clinical teaching staff had more knowledge and experiences in curricula integration, new teaching methods, assessment methods, and educational technologies. In the current Chinese medicine and health education context, curricula integration and the introduction of Problem-Based Learning (PBL) methods mainly takes place in the clinical medicine education stage of the education (normally year 3–5). Whether basic studies in year 2 and 3 should be integrated and whether PBL should be implemented in the earliest years of medicine and health education continue to be two debates in China (Sun, 2005; Sun, *et al.*, 2008; Liu, 2009). This explains why Clinical teaching staff members are more familiar with these aspects than Basic studies teaching staff, who are mainly responsible for teaching in year 2–3 and Public health teaching staff, whose teaching obligations are not part of clinical medicine education.

In the educational change process in China, a top-town approach is normally taken, meaning decisions and instructions come from the government, and administrators are the first group to receive training. This explains why Teaching administrators have a higher level of familiarity with the global change trends, international standards, and new policies. They also show a higher level of urgency towards reforms and change than other groups.

The University's geographical location (whether it was in Eastern or Western China) also had an influence on the research results. While there was a universal positive attitude towards change, universities in Western China reported a lower degree of knowledge and participation in most of the items discussed in the questionnaire, in particular, experiencing new teaching methods and assessment methods and access to educational technologies. This indicates a serious lack of resources in supporting reforms in the Western areas of China.

Teaching staff in this study generally held positive attitudes towards the importance of educational research, and the general development of medicine and health education. However, they have limited familiarity with the new trend of change occurring within medicine and health education both nationwide and at a global level. The general knowledge about these trends and standards are mainly restricted to certain groups of teaching staff. In addition, participation in educational research has been predominantly limited to a small group of teaching staff. This indicates the limitations of the institutional support that is extended; an improvement here could help to encourage and facilitate further educational reform.

10.1.4 Institutional Supports for Further Reforms

In order to be able to conduct and sustain reforms, the majority of the participants reported the need for institutional support in such aspects as training for new teaching and assessment methods and training in the use of technological resources. These findings also suggest that for reforms to be carried forward successfully, there is a general need for better distribution of educational resources.

In addition, the research findings indicate that although there is an overall positive situation with regards to institutional support for educational research in medicine and health, the current involvement is limited to certain groups of teaching staff. For further reform, there is a need to provide sufficient support at an institutional level to all groups of teaching staff, ensuring that they understand the reforms and have the necessary skills to implement the innovative methods. This finding to a great extent echoes suggestions from previous studies on educational reforms (Sheets and Schwenk, 1990; Mennin and Krackov, 1998; Carraccio *et al.*, 2002; Steinert and Mann, 2006; Steinert *et al.*, 2006).

10.2 Reflection on Teaching Staff's Motivation and Educational Change

Chapter 2 discussed a number of approaches to change at an organizational level. Previous work also suggested that educational innovation and change is rather complex and contextualized (Mennin and Krackov, 1998; Fullan, 2007; Du and Kirkebæk, 2012). Du (2011) studied whether innovative teaching and learning methods such as PBL can create a supportive learning environment, and in what ways this can take place. In her work, she developed a model to understand and analyze the culture of learning, as well as the change that happens in the learning process from three levels: the individual level, the institutional level, and the overall socio-cultural level. In relation to this study, a cultural change at each of these three levels has significance in the process of implementing educational reforms.

Educational change depends on what teachers do and on what they think (Fullan, 2007). Thus, individual members of teaching staff play an important role in the change process in education. At the individual level, participating in educational change involves new knowledge and skills for individual teaching staff and students (in this study we focus on teaching staff), and guidance for the practice of reflection throughout the change process. It is essential that the individual members of teaching staff are open to challenges and willing to engage in a professional and personal transformation process. The teaching staff's deep understanding of the innovations and reforms are of importance in building beliefs and commitment to the change process. Results in this study show that teaching staff members in medicine and health education in China have positive attitudes and the openness to innovation and change. However, they do not possess sufficient knowledge and the relevant skills that are needed for implementing change.

Educational reforms take place within an institutional context. Implementation of innovations and reforms will both improve the institutional profile and demand an institutional approach to support the implementation. Accordingly, it involves a series of considerations and actions in the aspects of institutional strategies, policies, regulations, physical facilities, curricula development, evaluation and quality assurance mechanisms, and the establishment of a new institutional culture that supports change. Previous experiences from reforms that were conducted in medicine and health education also showed that staff development does not only offer teaching staff knowledge and skills about alternative educational approaches, but also play a political role by empowering them as they learn about educational principles and current educational trends (Mennin and Krackov, 1998). All in all, the most important factor is to motivate teaching staff to get engaged in the change process (Dolmans and Wolfhagen, 1994; Maudsley, 1999; Cavanaugh, 2001; Kotter and Cohen, 2002; Savin-Baden, 2003; Moesby, 2004; Cohen, 2005; Malicky *et al.* , 2007; Fullan, 2007; Kotter, 2008). To motivate teaching staff's commitment, it is important to facilitate institutionalized staff development activities (for concrete recommendations on this, see the following section). In relation to this study, we can see that despite the institutional request, there is, so far, a general lack of support for motivating teaching staff's engagement in change.

Change does not only happen among individuals within the institution. The social-cultural context plays an important role in defining the process and effect of educational reform. Educational history, system, and overall beliefs (such as Confucianism as discussed in Chapter 2) can impact and restrict teaching staff's action in the change process. At the same time, changes in individuals and institutions will consequently create overall societal and cultural changes. Educational changes are driven by social needs and professional development, and at the same time, the changing of individual teaching staff, students, and medical universities will also create a new ideology of being a doctor and health professional in the Chinese society.

We see that the three levels of learning and change are intertwined and inseparable. In his study on educational change, Fullan (2007) emphasizes the importance of the intimate relation between staff and organization in the change process. He states that the classroom becomes effective when (1) quality people are recruited to teach and (2) the workplace is organized to energize teachers and reward accomplishments. In addition, overall, professional development and change of values in the society impact both individual and institutional change. In this way, individual change is linked with broader institutional and cultural practices. Personal, institutional and social betterment are intimately interconnected, as Fullan (2007) argues that large-scale change cannot be achieved if teachers, administrators, and schools only identify with their own progress and are not similarly connected with the success of others.

A sense of urgency has been highlighted in many studies and theories about initiating change at the institutional level (Kotter and Cohen, 2002; Moesby, 2004; Cohen, 2005; Kotter, 2008). It is important to help staff understand the gap between drawbacks and goals, and realize the necessity for change within the institution. In this study, a strong sense of urgency for reforms was observed at the three levels. Nationwide, most universities are experimenting with change in accordance with government guidelines. Within the universities, both administrators and teaching staff have agreed on the need for reforms. This indicates a positive sign of readiness for change at both the individual and the institutional level in medicine and health education in China. However, based on the research results of this study, what has been missing is the interconnection between the individuals and institutions. It is necessary to provide the teaching staff with a deep understanding of the issues, sufficient knowledge, and relevant skills.

A number of approaches to implement changes were discussed in Chapter 2. Drivers for educational change in China have so far been focused on government decisions and guidance, which is a top-down approach to change. Results from this study also evidenced a strong readiness for a bottom-up approach, with the strong overriding feeling being that there are positive attitudes towards change, and that change is needed among teaching staff in medicine and health education. We see this readiness as a positive sign for an interaction of both top-down and bottom-up approaches that will maximize the outcome of educational change. A normative-re-educative approach (Chin and Benne, 1985), as discussed in Chapter 2, can be recommended as a tool for further institutional development in order to facilitate the teaching staff's conception of change, as well as a process of consensus on the change.

To summarize, this interconnection perspective helps us to see what is taken for granted. In relation to educational practice, this view will improve understanding of how change takes place in a social and cultural context, and better facilitate a positive change process so as to maximize the outcomes.

10.3 Recommendations for Staff Development

Staff development remains an essential aspect in all educational developments. It plays an important role in the change process in pedagogy, didactics and management (Dolmans and Wolfhagen, 1994). For individual staff members, implementing innovation in their teaching practice involves complex competencies including knowledge, skills, awareness, engagement, and personal commitment (Maudsley, 1999; Savin-Baden, 2003). It is important that staff members have the motivation to implement innovative pedagogical methods by analyzing advantages and disadvantages of various options based on what has been made available to them (Malicky *et al.*, 2007). Cavanaugh (2001) suggests that the best approach for designing incentives for staff members is to have a range of options that can be tailored by individuals. As Fullan (2007) suggests, the solution to motivating people is to establish the right blend of tightness and looseness, or to build both into the interactive culture of the organization. Cooke and colleagues (Cooke *et al.*, 2010) proposed four goals of change in medical education, which may give inspiration for the contents design of staff development activities. These goals include: (1) standardization and individualization, (2) integration, (3) insistence on excellence, (4) formation of professional identity.

With these suggestions as inspiration to continue with the reflective evaluation on change and reforms, and based on results of this investigation, a list of recommendations is summarized for the improvement and development of teaching staff. These recommendations are aimed at educational institutions to facilitate staff development and further enhance educational reforms:

(1) For the recruitment of teaching staff, a certain number of teaching skill is required from the candidate, as opposed to only focusing on relevant educational background and research experience in medicine and health. It is recommended that the teaching staff recruited have an interest in teaching, even a passion for it, as well as openness to new and innovative ways of teaching.
(2) Provide pedagogy development activities for newly employed teaching staff before they start teaching. Activities can be short-term pedagogy programs where they are taught learning theories and develop teaching techniques. Mentorship can be established during their first academic year of teaching. For example, inexperienced teaching staff can be allocated a senior teaching staff working as a mentor to observe the teaching practices and give comments and suggestions.
(3) Provide ongoing training activities in order for all teaching staff to participate in high-quality professional development so they can practice and enhance student learning. These can be both short-term activities such as conferences, seminars, and workshops, and long-term activities such as participation in a semester-long, or longer, staff development program.
(4) Include diverse teaching evaluation methods, not only by management and students, but also peer-learning systems, to ensure that all teachers possess the skills and knowledge to improve student learning.
(5) Include teaching staff in the curricula design and integration process to ensure their understanding of the curricula and to encourage

their active participation in the design thereof. It is expected that as teaching staff feel ownership for the curricula, motivation for implementing innovation and reform will also increase.

(6) Engage teaching staff in the implementation of innovative methods in teaching and assessing students. A positive awards system and/or a possible reduction of other workload can help motivate staff engage with these implementations.

(7) Provide teaching staff with access to physical facilities such as educational technologies in order to better facilitate students learning more effectively, both from the perspective of technological society and globalized context.

(8) Encourage teaching staff to devote themselves to educational research, with supporting policies; for example, recognize their publications on educational research as part of their research achievements, and encourage teaching staff to participate in pedagogy conferences and seminars. This will involve facilitating a reduction of other workloads. It is also beneficial for both individuals and institutions to engage teaching staff to collaborate, not only with colleagues from the same medicine and health background, but also with colleagues with pedagogy and educational research backgrounds.

(9) Update teaching staff with knowledge of national and international standards in relation to student learning outcomes.

(10) Facilitate collaboration and effective communication within an institution, because they are necessary ingredients for meaningful and lasting changes (Mennin and Krackov, 1998).

10.4 Conclusions, Limitations and Future Perspectives

This study investigated teaching staff's attitudes toward, and participation in educational reforms in medicine and health in China. A nation-wide survey was conducted among 1955 teaching staff in 22 medical universities in China. Findings were analyzed and discussed in relation to theories on change in an international context. Results of this study evidenced an overall positive attitude toward educational reforms and participation at a certain level. However, teaching staff in medicine and health education in China have inadequate knowledge about the ultimate goals of the reforms, as well as limited understanding of the philosophy of student-centered learning concepts, which restricts their skills for implementing the actual innovations and reforms (Ludnerer, 2004). In addition, with limited participation in educational research, it is difficult for them to reflect on the meanings of change. This also limits the effect of reforms and change, particularly in relation to the goal of maximizing students' learning outcomes and competency development.

The empirical evidence of this study was mainly based on an investigation among teaching staff from medical universities in China. A quantitative method was used, aiming to gain an overview of teaching staff's attitudes toward and participation in educational reforms. It did, however, limit the possibility of in-depth understanding of teaching staff's perceptions and life experiences, which are very important aspects in the study of educational reforms. The results and limitations of the current study also suggest a series of considerations for further research topics; for example, qualitative research on teaching staff's life experiences in the change process, the role of leadership, the effect of staff development, and evaluation of change effects are also aspects that demand further attention with regard to understanding educational change in the Chinese context. In addition, cross-profession comparative studies on change, between medicine and health education for example, and other fields of professional education, would be a beneficial contribution. Cross-country studies would also add value to this field of research.

To conclude, this study suggests that sustainable educational change demands both top-down support and bottom-up engagement. Active participation from teaching staff is essential and institutional engagement is crucial. In order for the implementation of changes to be successful, two factors are essential from the teaching staff's perspective. Firstly, it is important for teaching staff to have a deep understanding of educational reform and change, and secondly, they should have sufficient knowledge and appropriate skills to be able to conduct the reforms through their teaching practices. To provide these two factors, institutional facilitation is necessary and vital. Therefore, institutionalized staff development is required urgently to further introduce innovation in the educational reform process. It is, and continues to be, essential to focus on methods that support teaching staff members in gaining knowledge and skills, and in their personal development process, on the journey of educational reforms and change.

References

[1] Accreditation Council for Graduate Medical Education (ACGME). *ACGME Core Competencies.* 1999. Available at http://www.acgme.org/acgmeweb/ . (Accessed Oct 2012).

[2] Alarcao, I. Changing to Project-Based Learning. The Role of Institutional Leadership and Faculty Development. *Management of Change: Implementation of Problem-Based and Project-Based Learning in Engineering,* E. de Graaff and A. Kolmos, (eds), SENSE Publisher, Rotterdam. 2007, pp. 69–83.

[3] Bloom, S.W. The Medical School as a Social Organization: the Sources of Resistance to Change. *Med Educ.* 23, 1989, pp. 228–241.

[4] Carraccio C, Wolfsthal SD, Englander R, Ferentz K, Martin C. Shifting Paradigms: From Flexner to Competencies. *Acad Med* 77(5), 2002, pp. 361–367.

[5] Cantor, J. C., Cohen, A. B., Barker, D.C., Shuster, A.L., and Reynolds, R.C. Medical Educators' Views on Medical Education Reform. *JAMA.* Vol. 265, No. 8, 1991, pp. 1002–1006.

[6] Cavanaugh, J.C. Make it So: Administrative Support for Problem-based Learning. *The Power of Problem-based Learning*, Duch, B. J., Groh, S. E. and Allen, D. E. (eds). Stylus, Sterling, Virginia. 2001, pp. 27–36.

[7] Chin, R., and Benne, K.D. General Strategies for Effecting Changes in Human Systems. In Bennis, W.G., Benne, K.D. and Chin, R. (Eds.). *The Planning of Change (4th edition).* New York: Holt, Rinehart and Winston. 1985, pp. 347–359.

[8] Cohen, D. *The Heart of Change Field Guide: Tools and Tactics for Leading Change in Your Organization.* Massachusetts: Harvard Business School Press, 2005.

[9] Cooke, M., Irby, D.M., and O'Brien, B.C., *Educating Physician: A Call for Reform of Medical School and Residency.* San Francisco: Jossey-Bass, Carnegie Foundation for the Advancement of Teaching, 2010.

[10] Cumming, A. and Ross, M. *The Tuning Project (Medicine) — Learning Outcomes/competences for Undergraduate Medical Education in Europe.* 2008. Edinburgh: The University of Edinburgh. Available at http://www.tuning-medicine.com/pdf/booklet.pdf (Accessed Oct 2012)

[11] Dolmans, D. and Wolfhagen, I. Improving the Effectiveness of Tutors in Problem-based Learning. *Medical Teacher*, Vol. 16, Issue 4, 1994, pp. 369–378.

[12] Du, X.Y. and Kirkebæk, M.J. Contextualising Task-Based PBL. In: Du, Xiangyun and Mads Jakob Kirkebæk (eds) *Exploring Task-Based PBL in Chinese Teaching and Learning.* Newcastle: Cambridge Scholars Publishing, 2012, pp. 172–184.

[13] Du, XiangYun. *Gender and diversity in a problem and project based learning* Aalborg: River Publishers, 2011.

[14] Frenay, M., Galand, B., Milgorm, E., and Raucent, B. Project- and Problem-Based Learning in the Engineering Curriculum at the University of Louvain. *Management of Change: Implementation of Problem-Based and Project-Based Learning in Engineering*, E. de Graaff and A. Kolmos, (eds), SENSE Publisher, Rotterdam. 2007, pp. 93–108.

[15] Fullan, M. *The New Meaning of Educational Change* (4th ed). New York: Teachers College Press, 2007.

[16] General Medical Council (GMC). *Tomorrow's Doctors — Outcomes and Standards for Undergraduate Medical Education*, 2009. Available at http://www.gmcuk.org/TomorrowsDoctors_2009.pdf_39260971.pdf (Accessed Oct 2012).

[17] IIME Core Committee, Institute for International Medical Education. *Global Minimum Essential Requirements.* 2002. Available at http://www.iime.org/documents/gmer.htm (accessed Oct 2012).

[18] Kotter, J.P. *A Sense of Urgency*. Boston, Massachusetts: Harvard Business Review Press, 2008.

[19] Kotter, J.P. and Cohen, D. *The Heart of Change – Real Life Stories of How People Change Their Organizations*. Boston, Massachusetts: Harvard Business Review Press, 2002.

[20] Liu, Q., *Research on the practice of PBL in medical education in China*. PhD thesis. Haerbin: Haerbin Medical University, 2009 (in Chinese).

[21] Ludmerer KM. Learner-centered medical education. *N Engl J Med*, 351, 2004, pp. 1163-1164.

[22] Malicky, D. M., Lord S. M., and Huang, M. Z. A Design Methodology for Choosing an Optimal Pedagogy: the Pedagogy Decision Matrix. *International Journal of Engineering Education*, Vol. 23, No. 2, 2007. pp. 325–337.

[23] Maudsley, G. Roles and responsibilities of the problem based learning tutor in the undergraduate medical curriculum, *British Medical Journal*, March 6, 318, 1999, pp. 657–661.

[24] Mennin, S. P. and Krackov, S. K. Reflections on Relevance, Resistance, and Reform in Medical Education. *Acad Med,* 73(9 Supp. l), 1998, pp. 60–64.

[25] Ministry of Education in China (MOE) *The Medical Education Criteria for Undergraduates*, 2005 (in Chinese).

[26] Moesby, E. (2004). Reflections on Making a Change towards Project Oriented and Problem Based Learning. *World Transactions on Engineering and Technology Education*, 3(2): 269–278.

[27] Savin-Baden, Maggi. 2003. *Facilitating problem-based learning: Illuminating perspectives.* Berkshire: Society for Research into Higher Education & Open University Press.

[28] Sheets KJ. and Schwenk TL. Faculty Development for Family Medicine Educators: An Agenda for Future Activities. *Teach Learn Med,* 2, 1990, pp. 141–148.

[29] Steinert Y. and Mann KV. Faculty Development: Principles and Practices. *J Vet Med Educ,* 33(3), 2006, pp. 317–324.

[30] Steinert Y, Mann K, Centeno A, Dolmans D, Spencer J, Gelula M, Prideaux D. A Systematic Review of Faculty Development Initiatives Designed to Improve Teaching Effectiveness in Medical Education: BEME Guide No. 8. *Med Teach* 28(6), 2006, pp. 497–526.

[31] Stojcevski, A., Du, X.Y., and Benz, T. Educational Innovation and Change for PBL. In: *Project Approaches to Learning in Engineering Education — The Practice of Teamwork*, Luiz Carlos De Campos, Ely Antonio Tadeu Dirani, Ana Lucia Manrique and Natascha van Hattum-Janssen (eds). Rotterdam, SENSE Publisher, 2012, pp. 77-88.

[32] Sun, B. Z. Trend for Educational Reforms in Medicine and Health. In: *Handbook for Applied Medical Pedagogy*, Sun, Baozhi and Zhao, Yuhong (eds). Beijing: People's Medical Publishing House, 2011, pp. 23–32 (in Chinese).

[33] Sun, B.Z., Liu, Q. & Zuo, T.M., PBL practice in CMU. *China Medical Education Research*, 7(9), 2008. pp. 1009–1011 (in Chinese).

[34] Sun, B. The New Objectives of Medical Education for the 21st Century and the Reform of Higher Medical Education. *Medical Science and Philosophy*, 26(2), 2005, pp. 24–26 (in Chinese).

Appendix 1

Survey on the Need and Current Situation of Medicine and Health Education Reform in China

Name: ______________________________

Gender: ______________________________

Age: ______________________________

Major/Profession: ______________________________

Job title: ______________________________

Position: ______________________________

University: ______________________________

Affiliation: ______________________________

Questions	Response
Q1, To what extent do you think teaching and learning methods and curricula design are important parts of medicine and health education?	Please rate 1–5, where 1 means strongly agree and 5 means strongly disagree
Q2, Which aspects of medicine and health education outcome should be further emphasized based on current indicators? Item 021, Practice-based learning and improvement Item 022, Interpersonal communication skills Item 023, Professionalism Item 024, Diagnosis and treatment of patients Item 025, The mastery of medical knowledge	Choose all that apply

Change and Reform in Medicine and Health Education in China, 189–198.

Questions	Response
Q3, To what extent do you think it is necessary for students in earlier study periods to access clinical courses such as Introduction to Clinical Medicine?	1 strongly agree to 5 strongly disagree
Q4, Which of the following aspects should be improved in medicine and health education based on the current situation?	Choose all that apply
Item 041, The connection between different parts of knowledge	
Item 042, Curricula integration	
Item 043, Emphasis on both teaching and learning	
Item 044, Knowledge memorization	
Q5, Has your subject been involved in the overall curricula integration?	Yes/No
Q6, To what extent do you believe that curricula integration should involve multiple subjects?	1 strongly agree to 5 strongly disagree
Q7, To what extent do you believe the traditional teaching method in medicine and health education should be reformed?	1 strongly agree to 5 strongly disagree
Or How necessary do you believe it is to reform the traditional teaching method in medicine and health education?	
Q8, To what extent do you think alternative teaching methods (e.g. PBL) should be introduced to medicine and health education?	1 strongly agree to 5 strongly disagree
Q9, Has PBL been implemented at your university?	Yes/No
Q10, What do you think of the effect of PBL?	1 very poor to 5 very good
Q11, Are you familiar with the concepts of mentoring and student support?	Yes/No
Q12, Are you familiar with the concepts of reflection and reflective practice?	Yes/No
Q13, Do you use reflective learning methods in your teaching?	Yes/No
Q14, To what extent do you agree that reflective learning methods should be implemented in medical teaching?	1 strongly agree to 5 strongly disagree
Q15, Are you familiar with the concept of lifelong learning?	Yes/No
Q16, To what extent do you think that the concept of lifelong learning should be integrated in the teaching and learning process?	1 strongly agree to 5 strongly disagree
Q17, Are you familiar with the concept of co-operative learning?	Yes/No
Q18, Are you familiar with the concept of portfolio building?	Yes/No
Q19, Are you satisfied with the wireless network on campus?	1 strongly agree to 5 strongly disagree
Q20, Do you have access to the wireless network on campus?	Yes/No
Is there campus wide wireless network in your university?	Yes/No
Q21, To what extent do you think that computer-based multimedia classrooms can meet the needs of teaching?	1 strongly agree to 5 strongly disagree
Q22, To what extent do you think that your university provides sufficient online teaching resources.	1 strongly agree to 5 strongly disagree
Q23, To what extent do you think that there are well-functioning distance education systems and resources at your university?	1 strongly agree to 5 strongly disagree
Q24, Which of the following educational technologies are accessible in the teaching process at your university?	Choose all that apply
Item 241, Projector and projection	
Item 242, Computer-based multimedia	
Item 243, TV and video	

Questions	Response
Item 244, Radio and recording	
Item 245, Virtual reality	
Item 246, Online teaching	
Q25, To what extent are you satisfied with the database retrieval system at your university?	1 strongly agree to 5 strongly disagree
Q26, Which of the following aspects educational technology training do you expect your university to provide?	Choose all that apply
Item 261, Computer basis	
Item 262, Office software	
Item 263, Use of internet	
Item 264, Database management	
Item 265, PPT making	
Item 266, Online teaching techniques	
Item 267, Online teaching management	
Item 268, Online teaching technology	
Q27, To what extent do you agree that teaching evaluations are an important part of the teaching and learning process?	1 strongly agree to 5 strongly disagree
Q28, Which methods should be used to evaluate teaching?	Choose all that apply
Item 281, Evaluation by management staff	
Item 282, Evaluation by peers	
Item 283, Evaluation by one's self	
Item 284, Evaluation by students	
Q29, Which kinds of evaluation methods have been used at your university?	Choose all that apply
Item 291, Evaluation by management staff	
Item 292, Evaluation by peers	
Item 293, Evaluation by selves	
Item 294, Evaluation by students	
Q30, Does your university use a computer-based medical test database for student assessment?	Yes/ No/Do not know
Q31, To what extent do you think there should be an assessment of basic medical knowledge for students who have taken the relevant courses after their 2nd and 3rd years of study?	1 strongly agree to 5 strongly disagree
Q32, Does your university organize examinations on basic medical knowledge?	Yes/ No
Q33, To what extent do you believe there should be an assessment of medical practice before students graduate?	1 strongly agree to 5 strongly disagree
Q34, Does your university organize an examination on medical practice before students graduate?	Yes/No
Q35, Are you familiar with the Standard Patient (SP) exam?	Yes/No
Q36, To what extent do you believe that the Standard Patient (SP) exam should be used on the medical practice assessment for graduation?	1 strongly agree to 5 strongly disagree
Q37, Does your university organize Standard Patient (SP) exams?	Yes/ No
Q38, Are you familiar with computer-based case simulation (CCS) examination?	Yes/ No
Q39, Does your university use computer-based case simulation (CCS) examinations?	Yes/ No

Questions	Response
Q40, To what extent do you think that exam results should be analyzed centrally and sent back to relevant departments afterwards?	1 strongly agree to 5 strongly disagree
Q41, Does your university analyze exam results centrally and send them back to relevant departments afterwards?	Yes/No
Q42, Are you familiar with portfolio assessment?	Yes/ No
Q43, Does your university have a unit focused on medicine and health education research?	Yes/No
Q44, To what extent do you agree with the necessity of establishing a unit focused on medicine and health education research?	1 strongly agree to 5 strongly disagree
Q45, Have you been or are you currently involved in any educational research projects?	Yes/No
Q46, If you replied yes to Q45, what levels of research projects were they?	Choose all that apply
Item 461, National	
Item 462, Provincial	
Item 463, Municipal	
Item 464, University	
Q47, How many articles on medical teaching and learning research have you published?	None/1–5/6–10/more than 10
Q48, To what extent do you think it is important to include professional morality, attitude, behavior, and ethics in medicine and health education?	1 strongly agree to 5 strongly disagree
Q49, To what extent do you think it is necessary to provide specific course and training activities on how to develop medical communication skills?	1 strongly agree to 5 strongly disagree
Q50, To what extent do you think it is important for medicine graduates-to-be to have knowledge of ethics and legislation issues in the profession?	1 strongly agree to 5 strongly disagree
Q51, To what extent do you think medical informatics courses should be provided in clinical medicine program.	1 strongly agree to 5 strongly disagree
Q52, To what extent do you think evidence-based medicine courses should be provided in clinical medicine program?	1 strongly agree to 5 strongly disagree
Q53, Are you familiar with the Global Standards of Undergraduate Medicine and health education?	Yes/No
Q54, Are you familiar with the accreditation of clinical medicine program conducted by the Ministry of Education?	Yes/No
Q55, To what extent do you think that the current medicine curricula in China should be reformed?	1 strongly agree to 5 strongly disagree
Q56, To what extent do you believe that the current medicine curricula contents should be updated?	1 strongly agree to 5 strongly disagree
Q57, To what extent do you think that the current continued education system for medical practitioners should be improved?	1 strongly agree to 5 strongly disagree
Q58, To what extent are you satisfied with the current medicine and health education system in China?	1 very satisfied to 5 very unsatisfied

Appendix 2

Higher Education System in China

In general, the Chinese education system (see the figure below) is composed of several levels:

- **Preschool education** lasts for 1 or 2 years. Students enroll in preschool at the age of 4 or 5.
- **Elementary education** lasts for 5 or 6 years and the student enrollment age is 6 or 7.
- **Secondary education** is composed of junior and high middle schools.
    - **Junior middle school** education lasts for 3 or 4 years with student enrollment age being around age 12.
    - **Higher middle school** education can be divided into general high school and vocational high school. Each of them lasts for 3 years.

Higher education consists of three levels.

- **Bachelor level education** includes general college education and vocational college education. **General college** is responsible for bachelor-level education, which usually lasts for 4 or 5 years. **Vocational college** is aimed to provide 2.5 to 3 year vocational courses. It is a vice-bachelor level education. Students usually enroll in general or vocational college at the age of 18 or 19 years old.
- **Master level education** lasts for 2 to 3 years.
- **Ph.D. level education** lasts for 3 to 6 years.

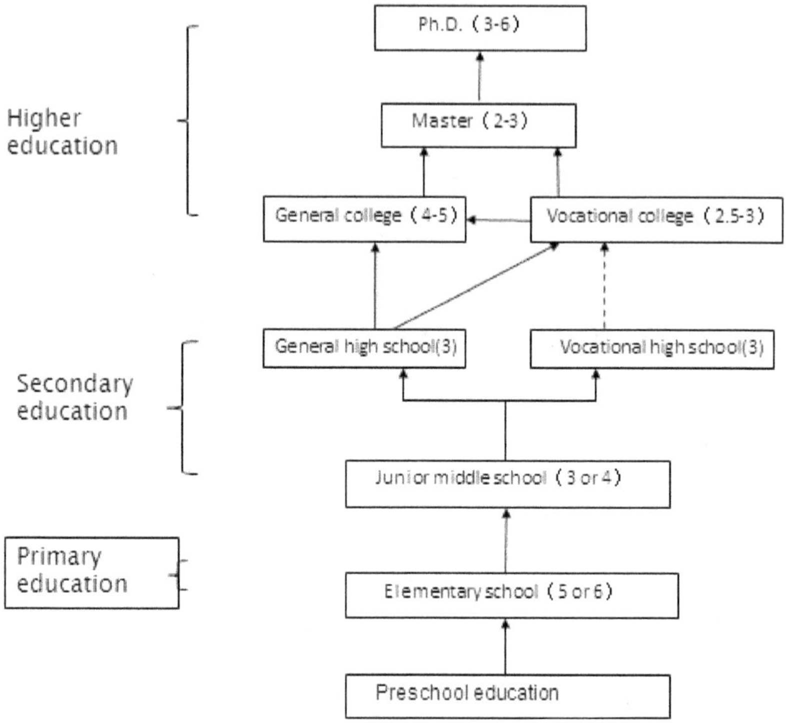

(The number indicates the duration of years)

Appendix 3

Higher Medical Education System in China

The degree program of higher medical education in China (see the figure below) is primarily composed of three paths, which students have to choose prior to their enrollment in the college. The first track includes a five-year bachelor program, three-year master program, as well as three-year Ph.D. program. The second track is a seven-year master program. After graduating from this program, students are directly granted a master's degree (without a bachelor's degree). Afterward, students can choose to either apply for a three-year Ph.D. program, or move directly to seeking employment. These two paths are the main higher education medical programs in China. Currently, there are a few medical schools, such as the Medical School of Peking University, that are experimenting with an eight-year Ph.D. program in which students are fostered as a Ph.D. student after their enrollment in the college.

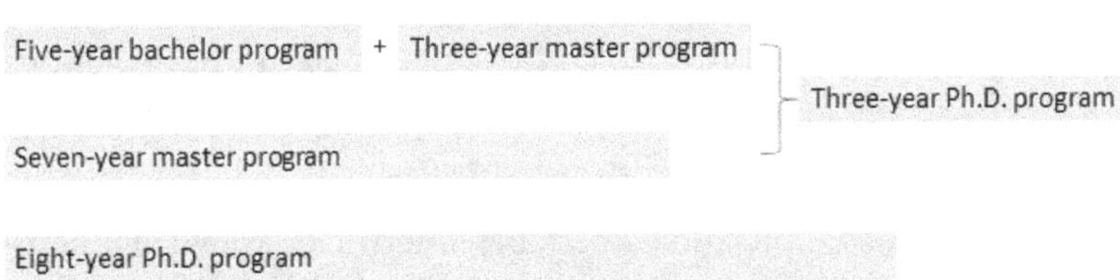

Appendix 4

PBL Curriculum Models at China Medical University

Currently, China Medical University is implementing PBL across the entire institution and thus formulated several PBL curriculum models (see the figure below). Mostly, instead of restructuring the entire curriculum system, PBL is established on the basis of traditional disciplinary subjects. Three major types of PBL models can be recognized:

- PBL as whole course approach: there is no lecture component for such a course; the whole course is delivered in the form of PBL activities. There are few tutorials.
- Partial PBL approach: a particular proportion of the knowledge content is delivered in the form of PBL. Depending on the function of PBL, this approach can be further divided into two sub-forms:
    - PBL as delivery approach: a certain section of the knowledge content is selected to be delivered in PBL. There is no lecture for this section; the knowledge content is totally delivered through PBL activities.
    - PBL as application: a certain section of the knowledge content is selected to be delivered in PBL. Lectures regarding such section are given beforehand, followed by a PBL unit in which students could apply the knowledge from the lecture in a PBL context. The aim of this PBL approach is to facilitate students' deeper understanding of the knowledge content that they learn from the lecture.

- PBL as a cross-course approach: since there are several integrated courses, PBL in this course can be considered as a cross-course approach. However, the objective and the use of PBL vary considerably, depending on the individual staff member.

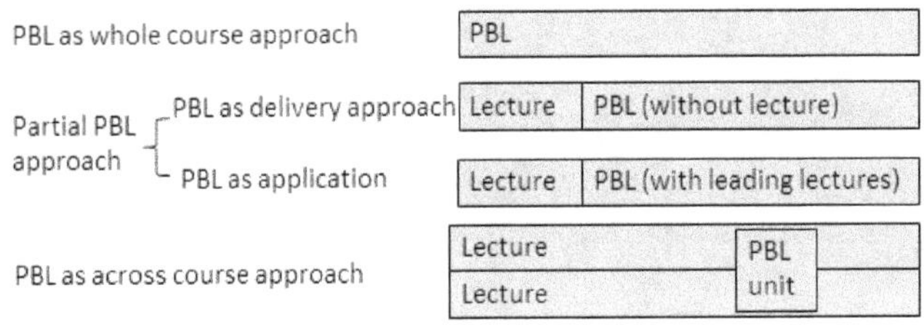

Index

access, 10, 30, 35, 42, 47–54, 67, 115, 117, 118, 124, 125, 134–139, 177, 178, 184, 190
accreditation, 3, 14, 24, 25, 29, 157, 158, 168–173, 177, 192
Accreditation Council for Graduate Medical Education (ACGME), 3
achievement, 19, 184
age, 39, 41–43, 46, 48–51, 56, 57, 62, 66, 68–70, 72–74, 79–81, 85, 86, 88–91, 96, 100, 101, 104, 108, 109, 113, 115, 119–121, 125, 126, 130–132, 135, 138, 142, 143, 146, 149, 152, 154, 155, 158, 160, 161, 170, 173, 176, 177, 193
analytical reasoning, 3
assessment, 2–8, 10, 13, 16, 18–22, 24, 26–28, 30, 33, 35, 40, 87–94, 103–115, 175–179, 191, 192
assessment methods, 4, 5, 8, 10, 16, 18, 21, 24, 27, 35, 40, 87, 115, 175–179
Association for Medical Education in Europe (AMEE), 3
Association of Medical Education in China, 6
attitudes, 6–10, 17, 34, 40, 42, 47, 49, 50, 67, 70, 71, 73, 85, 87, 92, 118, 122, 124, 157, 175–178, 180, 182, 184, 185
attributes, 3
authentic, 19–21, 33
average, 7, 151, 155

background, 1, 25, 39, 42, 46, 175, 183, 184
basic medical knowledge, 87–94, 105–113, 115, 191
basic studies, 9, 43, 44, 51, 52, 57, 58, 63, 66, 69, 75, 76, 82, 91–93, 97, 101, 104, 110, 121, 122, 126, 127, 131, 132, 135, 136, 138, 139, 143, 144, 147, 150, 152, 153, 155, 163–165, 171, 173, 174, 178
basic studies teaching staff, 9, 44, 51, 52, 57, 58, 63, 66, 69, 75, 76, 82, 91–93, 97, 101, 104, 109, 110, 121, 122, 126, 127, 131, 132, 135, 136, 138, 139, 143, 144, 147, 150, 152, 153, 155, 163–165, 171, 173, 174, 178
behavior, 29, 32, 34, 157–169, 172, 173, 192
bottom-up approach, 31, 182

change, 1, 2, 4, 7–10, 13–17, 19, 21–25, 27–35, 39, 40, 45, 50, 52, 54, 69, 70, 174–185
China medical board, 8, 40, 42
Chinese medical education, 4, 13, 24–26, 29, 35
clinical courses, 16, 26, 47–55, 67, 190
clinical education, 5
clinical medicine, 47–54, 67, 68, 157–173, 177, 178, 190, 192
clinical practice, 5
clinical reasoning, 3
clinical skills, 3, 5, 15, 19
clinical teaching staff, 9, 44, 51, 52, 57, 58, 63, 66, 69, 75, 76, 80–82, 85, 91–93, 97, 101, 104, 108–110, 114, 121, 122, 126, 127, 131, 132, 135, 136, 138, 139, 143, 144, 146, 147, 150, 152, 153, 155, 163–165, 171, 173, 177, 178
co-operative learning, 72, 78–85, 190
collaboration, 20, 42, 184

communication skills, 1, 3, 15, 27, 28, 30, 41, 55–60, 68, 69, 157–169, 172, 173, 176, 177, 189, 192
community, 1
competences, 15, 21, 69, 176
competency-based medical education, 33
Computer-Based Case Simulations (CCS), 20
conferences, 5–7, 32, 33, 183, 184
context, 182
contextual, 2, 6, 33
criteria, 6, 16, 21, 24, 26
critical thinking, 3, 15
curricula contents, 3, 158–162, 164, 165, 167–169, 192
curricula design, 4, 40, 47–54, 67, 68, 183, 189
curricula integration, 5, 8, 10, 16, 40, 41, 47–55, 60–69, 175–178, 190

database retrieval, 117–124, 137, 191
diagnose-treat model, 25
diagnosis and treatment of patients, 41, 55–60, 69, 189
didactic approach, 17, 18, 22, 31, 35
disease-centered, 2
distribution, 41–46, 139, 155, 179

east group, 54, 55, 60, 65, 67, 83–85, 94, 98, 102, 103, 105, 112, 114, 115, 124, 128, 133, 136, 138, 139, 144, 147, 148, 151, 153–155, 167
educational change, 9, 13, 21, 22, 24, 28, 29, 31–35, 50, 176, 178–182, 185
educational institutions, 4, 9, 13, 16, 22–24, 183
educational leaders, 7
educational outcomes, 2, 68, 69, 176
educational reforms, 6–8, 10, 26, 40, 42, 49, 54, 175, 176, 179, 180, 183–185
educational research, 5, 8, 10, 40, 141, 145–148, 154–156, 176–179, 184, 185, 192
educational technologies, 5, 8, 10, 16, 20, 35, 40, 117, 124–128, 138, 139, 175–178, 184, 190
educational technology training, 117, 118, 129–134, 137–139, 191
empirical basis, 46
encourage, 5, 17, 19, 34, 179, 183, 184
ethics, 3, 5, 25, 31, 157–169, 172, 173, 192
evaluation, 2, 5, 16, 87–103, 113–115, 180, 183, 185, 191

facilitate, 8, 16, 17, 19, 21, 23, 174, 179, 180, 182–184, 196
facilitator, 18, 27
familiarity, 41, 71, 78–86, 88, 105–113, 115, 157, 158, 168–174, 177, 178
Flexner report, 2, 14, 34
formative assessment, 27
framework, 2, 3, 15, 16

gender, 39, 41, 42, 46, 48, 49, 55, 56, 61, 66, 68, 69, 72, 73, 79, 80, 85, 88, 89, 95, 96, 99, 100, 103, 104, 107, 113, 115, 118, 119, 125, 126, 129, 130, 134, 135, 137, 142, 145, 146, 148, 149, 151, 154, 155, 158–160, 169, 170, 173, 176, 177
general development, 10, 157–162, 164, 165, 167–169, 172, 174, 178
general knowledge, 174, 179
General Medical Council (GMC), 3
geographical location of the university, 68–70, 86, 176
Global Minimum Essential Requirements (GMER), 3
Global Standards of Undergraduate Medical Education, 157, 158, 168, 170–173
goals of medicine, 1, 2
government, 4, 13, 22, 23, 29, 35, 178, 181, 182
graduates, 3–5, 16, 24, 29
graduates-to-be, 157–169, 172, 173, 192
group work, 18, 27, 34

hastings center report, 1
health sector, 1
healthcare service, 2
hermeneutic, 32, 33

implement, 33, 179, 182
individual level, 22, 179, 180

innovation, 5, 8, 10, 14, 17, 21–23, 28, 29, 33–35, 40, 71, 175, 177, 179, 180, 182, 184, 185
innovation practices, 5, 40
innovative methods, 179, 184
institutional level, 156, 179–181
institutional support, 8, 10, 13, 40, 141, 154, 156, 174, 175, 179
institutionalized support, 34, 139
integrated curricula, 5
integration, 3, 5, 8, 10, 13, 14, 16, 25, 26, 29, 30, 40, 41, 47–55, 60–69, 72, 73, 75, 76, 78, 175–178, 183, 190
international standards, 5, 24, 40, 178, 184
interpersonal skills, 3
introduction to clinical medicine, 47–54, 67, 190
investigation, 8, 9, 14, 16, 40–42, 44, 48, 68, 85, 86, 173, 175–177, 183, 185
involvement, 7, 8, 10, 40, 47–49, 51, 65, 71, 141, 145–148, 150, 154–156, 175, 177–179

job title, 39, 41, 44, 46, 48, 53, 54, 58, 59, 64, 67–70, 72, 76, 77, 81–83, 85, 86, 88, 92–94, 97, 98, 102, 105, 110, 111, 113, 115, 118, 122, 123, 127, 128, 132, 133, 136–139, 142, 144, 147, 150, 153–155, 158, 166–168, 171–174, 176–178, 189
junior group, 53, 59, 64, 82, 98, 102, 110, 127, 133, 138, 147, 150, 153, 155, 166, 171, 173, 174

knowledge, 3, 5–7, 14–18, 22–27, 30–35, 40, 41, 47, 55–65, 67, 69, 78–94, 105–113, 115, 124, 134, 141, 145, 154, 157–169, 172–174, 177–180, 182–185, 189–192, 196
knowledge memorization, 30, 60–65, 69, 190

learner-centered, 2
learning, 2, 3, 5, 6, 8, 10, 14–24, 26–31, 33–35, 40, 41, 47–65, 67–69, 71–85, 87–94, 113, 114, 141, 142, 151–154, 175–179, 181, 183–185, 189–192
learning outcomes, 3, 6, 14, 15, 18, 21, 177, 182, 184, 185
learning process, 179, 182
lecture-based, 5, 17, 18, 26, 27, 29
lifelong learning, 17, 71–85, 190
Likert-type questions, 41, 68, 84, 113, 137, 173
limitation, 10, 29, 30, 39, 41, 42, 46, 174, 177, 179, 184, 185

management, 3, 6, 23, 31, 33, 95–103, 114, 129–134, 138, 182, 183, 191
massification, 30, 35
mean, 48, 88, 118, 151, 155
medical curricula, 2
medical degree qualifications, 3
medical education criteria for undergraduates, 24, 26
medical knowledge, 3, 25, 41, 55–60, 87–94, 105–113, 115, 189, 191
medical practice, 3, 25, 28, 87–94, 105–113, 115, 191
medical school, 2–4, 9, 14, 16–19, 21, 22, 24–26, 28–30, 35, 44, 45, 195
medical school students, 3
medical teaching and learning research, 142, 151–154, 192
medical training, 3
medical universities, 4–6, 9, 24, 42, 85, 88, 105, 113, 175, 181, 184, 185
medicine and health education, 1–8, 10, 13, 33, 39, 40, 43, 47–57, 60–62, 65, 67–78, 84–90, 92, 113, 141–148, 154, 155, 157–169, 172–182, 184, 185, 189, 190, 192
mentoring, 71, 78–85, 190
methodology, 9, 17, 27
middle group, 53, 59, 64, 76, 82, 83, 92, 98, 102, 110, 111, 123, 124, 128, 133, 136, 138, 139, 147, 150, 153, 166, 171, 173
minimum learning outcomes, 3
ministry of education of China, 157, 172
ministry of health in China, 6
morality, 157–169, 172, 173, 192
motivate, 5, 23, 30, 32, 180, 184

motivation, 18, 20, 24, 28, 31–34, 179, 182, 184
multiple subjects, 47–55, 67, 68, 190
multiple-choice questions, 41, 68, 113, 137

nationwide, 6, 9, 14, 21, 24, 174, 179, 181
normative-re-educative approach, 32, 182
north China center for medical education, 6–8, 39, 40, 42

objective structured clinical examination (OSCE), 5, 18, 28
outcome-based approach, 2
overall opinion, 172

P-value, 50, 51, 53, 73, 75, 76, 90–92, 119, 121, 123, 143, 144, 152, 153, 161, 163, 166
participants, 6, 33, 39–48, 50, 51, 53–74, 76, 78–90, 95–106, 113–115, 117–119, 121, 122, 125–139, 141–145, 147–151, 154, 155, 157–160, 162, 163, 166, 168, 171–173, 176–179
patient care, 3
patient-centered, 2
patients, 3, 15, 18, 19, 25, 28, 34, 41, 55–60, 69, 189
pedagogy development, 6, 7, 16, 17, 39, 183
pedagogy innovation, 5, 8, 10, 40, 71, 175
pedagogy methods, 4
physical facilities, 16, 117–119, 121, 122, 124, 137, 139, 180, 184
physicians, 6, 8, 14
portfolio, 5, 19, 20, 72, 78–85, 106–112, 115, 190, 192
portfolio assessment, 106–113, 115, 191, 192
portfolio building, 72, 78–85, 190
positive attitudes, 49, 72, 73, 85, 92, 113, 118, 122, 124, 139, 142, 144, 154, 159, 160, 162, 166, 174, 176–178, 180, 182, 184
post hoc test, 50, 51, 53, 73, 75, 76, 90–92, 119, 121, 123, 143, 144, 152, 153, 161, 163, 166
power-coercive, 32
practice-based learning and improvement, 3
Problem and Project Based Learning environment, 186
problem situation, 17
problem solving, 17, 18, 26–28, 30
Problem-Based Learning, 5, 8, 16, 17, 40, 175, 178
profession, 1–3, 50, 51, 57, 58, 63, 64, 66, 75, 80, 81, 90, 91, 97, 101, 104, 108, 109, 114, 121, 126, 127, 131, 132, 135, 138, 139, 143, 144, 146, 149, 150, 152–155, 158, 162, 163, 171, 173, 185
professional values, 3, 6, 15
professionalism, 3, 7, 15, 41, 55–60, 68, 176, 189
public health teaching staff, 9, 44, 51, 52, 57, 58, 63, 66, 67, 75, 76, 81, 82, 91–93, 97, 101, 104, 108–110, 121, 122, 127, 131, 132, 136, 139, 143, 144, 147, 150, 152, 153, 155, 163–165, 171, 173, 178
publication, 141, 151–155, 184

quality assurance, 4, 24, 180
quality education, 23, 35
quality of education, 4
questionnaire, 8, 40–44, 55, 175, 178

recommendation, 10, 14, 21, 176, 180, 182, 183
reflection, 10, 17, 19–21, 72, 78–84, 179, 180, 190
reflective learning, 71–84, 190
reflective practice, 72, 78–84, 190
reform, 1, 2, 4–10, 13, 14, 16, 21, 22, 24–27, 29, 34, 39, 40, 42, 47–51, 53–55, 67, 68, 70, 85, 117, 159, 160, 163, 166, 167, 172, 173, 176–178, 184, 185, 189, 190
research project, 7, 39, 141, 148–151, 154, 155, 175, 192
research task force of quality assurance for Chinese medical education, 24
resources, 4, 30, 35, 45, 117–124, 137, 139, 155, 178, 179, 190

sample, 19, 43
scientific knowledge, 3
self-report, 41
seminars, 6, 7, 32, 33, 183, 184
senior group, 53, 58, 59, 64, 76, 82, 83, 92, 97, 98, 102, 111, 114, 123, 124, 127, 128, 133, 136, 138, 147, 150, 153, 155, 166, 171, 174
significant difference, 41, 44, 49–51, 53–57, 59–61, 65–67, 76, 79, 82, 85, 95, 97, 98, 102, 105, 108, 111, 112, 114, 127, 131, 136–138, 146–155, 160, 162, 166, 171–173, 177
single-choice question, 103, 113, 154
skills, 1–3, 5, 6, 15, 17–20, 23, 27, 28, 30, 32–34, 41, 55–60, 68, 69, 86, 115, 139, 157–169, 172, 173, 176, 177, 179, 180, 182, 183, 185, 189, 192
societal need, 4, 13, 15, 34, 35
staff/faculty development, 7, 18, 33, 34, 40
standard, 3, 5, 6, 24, 25, 35, 40, 89, 90, 93, 94, 157, 158, 168, 170–174, 177–179, 184
Standard Patients (SP), 5, 87–94, 105–113, 115, 191
standard patients exam, 89–94, 105–113, 115, 191
student directed learning, 30
student support, 71, 78–85, 190
student-centered learning, 5, 177, 185
students' performance, 6
systems-based practice, 3

teacher-centered, 5
teaching administrators, 7, 9, 43, 44, 51–53, 57, 58, 63, 66, 69, 75, 76, 80–82, 85, 91–93, 97, 101, 104, 108, 110, 114, 115, 121, 122, 127, 131, 132, 136, 139, 143, 144, 146, 147, 150, 152–155, 163–165, 171, 173, 177, 178
teaching and learning method, 27, 28, 33, 40, 47–52, 54, 67, 68, 179
teaching effect, 23
teaching evaluation, 87–94, 113–115, 183, 191
teaching practice, 8, 40, 138, 175, 176, 182, 183, 185
teaching staff, 4–10, 19, 33, 39, 40, 42, 44, 47, 51–53, 57, 58, 63, 66, 67, 69, 71, 72, 75, 76, 80–82, 85, 87, 91–93, 97, 101, 104, 108–110, 114, 117, 121, 122, 126, 127, 131, 132, 135, 136, 138, 139, 141, 143, 144, 146, 147, 150, 152–156, 158, 163–165, 171, 173–185
test, 18, 19, 21, 28, 41, 88, 103, 104
the core committee of the institute for international medical education (IIME Core Committee), 3
the tuning project (Medicine), 177
tomorrow's doctors, 3, 15, 34, 177
top-down approach, 31, 182
traditional teaching method, 47–55, 68, 190
transformation, 1, 2, 21, 33, 180
trend, 2, 13, 15, 26, 69, 96, 98, 138, 174, 178
tuning project, 15, 69, 177

understanding, 3, 16, 17, 19, 23, 32, 40, 177, 180, 182, 183, 185, 196
united states medical licensing examination (USMLE), 19
urgency, 31, 35, 178, 181

west group, 54, 55, 60, 64, 65, 67, 76, 83–85, 92, 94, 98, 102, 103, 105, 111, 112, 114, 124, 128, 133, 136, 138, 139, 144, 147, 148, 150, 151, 153–155, 166, 167, 171, 172
Western education system, 4
western medicine, 2, 4
wireless network, 117–124, 134–139, 190
workshop, 6, 32, 33, 183
world federation for medical education (WFME), 3
world health organization (WHO), 1
worldwide, 2, 4, 13, 16, 18, 174

Yes/No' questions, 41, 68, 78, 84, 113, 145, 154, 168

Change and Reform in Medicine and Health Education in China — A Teaching Staff's Perspective

Xiangyun Du, Jiannong Shi, Yuhong Zhao and Baozhi Sun (eds)

In recent decades, medicine and health education has been challenged worldwide by changes in its profession. Being a doctor nowadays encompasses much more than having biomedical knowledge and includes interdisciplinary skills related to societal needs, communication skills, and ethical consideration, among other things In order to provide these skills and competences, many medical schools are implementing changes in different aspects of the education.

These changes are also occurring in China. In the past twenty years medical education in China has initiated a series of reforms. The current reforms have mainly been led by the Ministry of Education and Ministry of Health. These initial actions have evidenced both positive and negative attitudes and reactions Is there a need to make further reforms and changes? If so, in what aspects? This book documents a national investigation of attitudes from teaching staff on the reforms and changes. Nearly 1800 teaching staff from 23 medical universities participated in this investigation. The results suggest that sustainable educational change demands not only supports from policy-makers and leaderships but also active participation from teaching staff. In order for the implementation of reforms and changes to be successful, two factors are essential from the teaching staff's perspective. First, it is important for teaching staff to gain a deep understanding of educational reform and change, and second, they should develop appropriate skills to be able to conduct the reforms through their teaching practice. To provide these two factors, institutional facilitation is necessary and crucial.

For Product Safety Concerns and Information please contact our EU representative GPSR@taylorandfrancis.com
Taylor & Francis Verlag GmbH, Kaufingerstraße 24, 80331 München, Germany

www.ingramcontent.com/pod-product-compliance
Lightning Source LLC
LaVergne TN
LVHW010629300926
849162LV00001B/98
9788792982346